ITALIAN PHYSICAL SOCIETY

PROCEEDINGS

OF THE

INTERNATIONAL SCHOOL OF PHYSICS
« ENRICO FERMI »

COURSE L

edited by J. COULOMB
and M. CAPUTO

*Varenna, Italy · Scuola internazionale di Fisica*

VARENNA ON LAKE COMO
VILLA MONASTERO
13th - 25th JULY 1970

# Mantle and Core
# in Planetary Physics

1971

*ACADEMIC PRESS* · *NEW YORK AND LONDON*

*B7/72*

SOCIETA' ITALIANA DI FISICA

RENDICONTI

DELLA

SCUOLA INTERNAZIONALE DI FISICA
« ENRICO FERMI »

L Corso

a cura di J. Coulomb
e M. Caputo

VARENNA SUL LAGO DI COMO
VILLA MONASTERO
13 - 25 LUGLIO 1970

# Mantello e nucleo nella fisica planetaria

1971

*ACADEMIC PRESS* • *NEW YORK AND LONDON*

ACADEMIC PRESS INC.
111 FIFTH AVENUE
NEW YORK 3, N. Y.


*United Kingdom Edition*
Published by
ACADEMIC PRESS INC. (LONDON) Ltd.
BERKELEY SQUARE HOUSE, LONDON W. 1

*Library of Congress Catalog Card Number: 70-190385*

PRINTED IN ITALY

# INDICE

# Préface.

J. Coulomb

*Faculté des Sciences - Paris*
*Directeur du Cours*

Les cours d'été dédiés à Enrico Fermi, que la Société Italienne de Physique organise chaque année à Varenna sur le lac de Côme, out acquis et maintenu depuis leur début en 1953 la brillante réputation souhaitée par leurs fondateurs. Consacrés à la physique moderne, ils ont rapidement fait une place à l'astrophysique, mais la géophysique était jusqu'ici restée à l'écart. Aussi lorsque, en février 1969, le Prof. Toraldo di Francia, Président de la Société Italienne de Physique, me demanda d'organiser, avec le concours du Prof. Caputo, un cours consacré à la Géophysique et à la Physique des Planètes, acceptai-je cette offre avec empressement. Il s'agissait des « noces d'or » de l'Ecole Internationale de Physique « Enrico Fermi » puisque ce cours devait être le cinquantième depuis sa fondation. Je pense qu'il n'a pas été indigne de ses devanciers.

L'organisation d'une réunion scientifique est toujours une petite aventure. Matériellement, cette aventure a été réduite au minimum grâce à la compétence et au dévouement du Prof. Germanà, secrétaire de la Société Italienne de Physique, grand ordonnateur des Cours. Cependant il y a d'autres complications possibles: les scientifiques se déplacent volontiers, mais des raisons universitaires, financières, politiques, et malheureusement aussi des raisons de santé peuvent modifier la composition du corps professoral, donc le contenu détaillé des sujets abordés. Je remercie ici ceux qui ont bien voulu remplacer au dernier moment les conférenciers empêchés, savoir le Prof. Lowes remplaçant le Prof. Runcorn, le Dr. Ahrens remplaçant le Dr. Keeler, et le Prof. Jacobs ajoutant à la tâche qui lui avait été dévolue celle qui revenait à Mme Lubimova.

Finalement la liste des conférenciers a été la suivante:

Dr. T. J. Ahrens
California Institute of Technology, Pasadena

Prof. M. Caputo
Istituto di Fisica dell'Università, Bologna

Prof. G. Colombo
Facoltà di Ingegneria, Università degli Studi, Padova

Prof. J. Coulomb
Faculté des Sciences, Paris

Prof. J. A. Jacobs
The University of Alberta, Alberta

Prof. V. Keilis-Borok
Institute of Physics of the Earth, Academy of Sciences of the USSR, Moscow

Prof. L. Knopoff
University of California, Los Angeles, Cal.

Prof. J. Lowes
School of Physics, The University, Newcastle upon Tyne,

Prof. W. V. R. Malkus
Department of Mathematics, M.I.T., Cambridge, Mass.

Prof. F. Press
Earth and Planetary Science Department, M.I.T., Cambridge, Mass.

Dr. L. Thomsen
Laboratoires des Hautes Pressions, Bellevue

Prof. R. O. Vicente
Facultade de Ciencias, Lisboa

Certains de nos étudiants étaient des physiciens et non des géophysiciens. Des conférences d'introduction leur ont donc été faites pour les mettre à même de profiter des conférences plus avancées qui les amèneraient à l'extrême pointe de la science. A notre grand regret il n'était guère possible de publier en volume la totalité des leçons offertes. Nous avons choisi celles qui nous ont paru, soit par leurs qualités didactiques, soit par l'originalité des recherches qui y sont décrites, pouvoir intéresser un large public. L'ensemble des conférences publiées est un peu hétérogène, et ne correspond pas exactement au titre du Cours puisque les planètes autres que la Terre n'y apparaissent guère; mais à l'exception d'un ou deux sujets particuliers le présent livre évoque assez bien l'ensemble présenté aux étudiants.

Je n'avais jamais visité la Villa Monastero. Je l'avais aperçue de loin, au cours d'un séjour à la Villa Serbelloni, de l'autre côté du lac, pour un colloque organisé par l'UNESCO et par la Fondation Rockfeller. C'est un lieu extraordinaire. Ceux qui liront cette préface ont certainement entendu parler des beautés du Lac de Côme, ne serait-ce que pour avoir traduit au lycée « quid agit Comum, tuae meaeque deliciae? ». S'ils n'ont pas été jusqu'à Varenna, ils peuvent difficilement se représenter la Villa Monastero enfouie dans les arbres et les fleurs odorantes. Si j'en parle ici, c'est pour leur recommander de ne pas manquer l'occasion d'y venir si la Géophysique a un jour la bonne fortune de bénéficier d'un nouveau cours d'été.

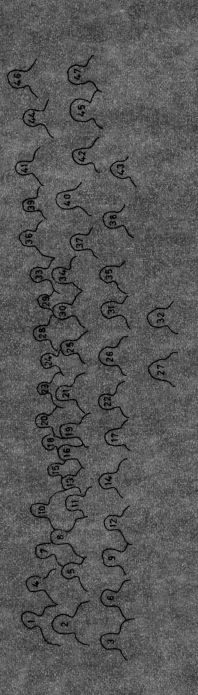

1. A. Bottari
2. E. Boschi
3. J. A. Jacobs
4. C. Denis
5. H. P. Harjes
6. W. V. R. Malkus
7. J. L. Le Mouël
8. J. Brown
9. F. Press
10. L. Battiston

11. G. Giunchi
12. V. Keilis-Borok
13. G. Molfchan
14. L. Knopoff
15. G. Calcagnile
16. A. Piva
17. G. Germanà
18. L. Secco
19. G. Panza
20. G. P. Gregori

21. D. Postpischi
22. L. Thomsen
23. C. Valenti
24. R. Console
25. L. Guarnieri-Botti
26. M. Caputo
27. N. Loperfido
28. F. Giorgetti
29. P. Pavese

30. M. C. Frazer
31. R. O. Vicente
32. P. Longhi
33. F. Mainardi
34. H. Neugebauer
35. J. Coulomb
36. D. Anfossi
37. B. M. Belli
38. G. Colombo

39. S. Barbarino
40. P. Baxa
41. J. Berger
42. A. Pennisi
43. M. T. Cinti
44. G. Napoleone
45. G. Losito
46. B. L. N. Kennet
47. M. T. Quagliarello

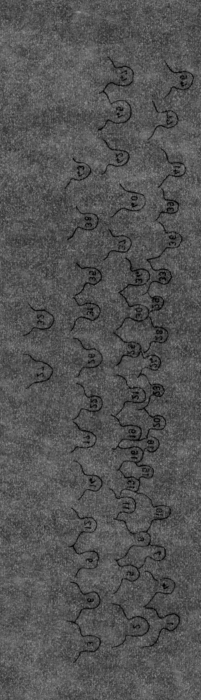

SOCIETÀ ITALIANA DI FISICA

SCUOLA INTERNAZIONALE DI FISICA « E. FERMI »

L CORSO - VARENNA SUL LAGO DI COMO - VILLA MONASTERO - 13 - 25 Luglio 1970

# A Model for Continental Drift.

W. V. R. MALKUS

*Department of Mathematics, M.I.T. - Cambridge, Mass.*

PART I

## 1. – Self-convection of line heat sources.

Geophysical literature of the past few years has been filled with fascinating observations on the drift of continents. The new global tectonics (ISACKS *et al.*, 1968) paints a picture of North and South America moving away from Africa and Europe at a rate of approximately four centimeters a year with significant upwelling of deep material in the mid-Atlantic ridge. This motion has forced the crust of the Pacific Ocean to turn down at the continental edges, producing the ring of earthquakes and vulcanism in the Pacific basin. The exploration of this global convection adds new understanding of the history of our earth every month. However, the dynamical basis for this motion has been investigated in only a casual fashion. The story told in the literature is of convective motions due to heat sources deep in the mantle of the Earth, moving the (passive) continents around; yet there is no evidence for an energy source at depth to provide for such convective processes.

The only well-established source of heat is the uranium and other radio-active materials in the continental masses themselves and a smaller amount in the oceanic basalt. A source of heating in the continental material on top of deeper, perhaps less viscous material, would have a stabilizing effect in the usual convective sense. However, as early as 1935 PEKERIS established that convection would exist due to an imposed horizontal temperature gradient, even though the underlying fluid were stable. PEKERIS did not discuss, nor at that time was it determined that continents were drifting, but he did suggest that temperature differences of 100 degrees or so between continental structures and oceanic structures would lead to fluid motions of about a centimeter a year. PEKERIS assumed certain smooth global distributions of temperature variation and computed the slow velocity fields which would result from these temperature distributions in a uniform fluid.

In this work we will explore the consequences of the assumption that the principal source of heat for continental motion is the distribution of radioactive materials in the continental and oceanic crust. Our models are attempts to isolate the simplest examples of motion induced by such horizontal inhomogeneity in heating and yet retain what we believe to be the features most essential for models which may have usefulness in the general interpretation of geophysical data. While we consequently have excluded the variation of viscosity with depth, the solidification of crustal material, and of course the geometric complexity of the real geophysical process, we can yet hope that if the significant dynamical aspects of the continental drifting phenomenon are contained in our idealization some of these complexities can be added at a later date.

Both models we contruct are approximately realizable in the laboratory, with the purpose in mind of testing the limits of validity of our theoretical proposals. The first model has the virtue that theoretical results for the entire field of motion can be attained. It has the disadvantage of mathematical complexity and un-geophysical appearance. We encourage the earth-scientist to bear with us until we discuss drifting block models in the second Part.

We consider two line heat sources of strength $Q$ per unit length in a fluid of depth $h$. The heat sources are constrained to move at a depth $d$ and at any instant are a distance $2a$ apart. The upper and lower bounding surfaces are isothermal, the lower one being held at a temperature $\Delta T$ above that of the upper. We consider two-dimensional motions induced by these heat sources, the stream function for such motion satisfying the condition that it and its second derivative with respect to the vertical co-ordinate ($z$) vanish at the upper and lower surface, *i.e.* rigid slippery boundaries. On the line of symmetry these two sources, the stream function and its second derivative with respect to the horizontal co-ordinate ($x$) vanish and the horizontal gradient of the temperature also vanishes. Figure 1 indicates one-half of this symmetric distribution.

As posed, this problem is well-suited for numerical computation even when the viscosity is a complicated function of temperature. However, we seek analytical solutions and prescribe the kinematic viscosity ($\nu$), the thermometric conductivity ($\varkappa$), and the coefficient of expansion ($\alpha$) of the fluid to be

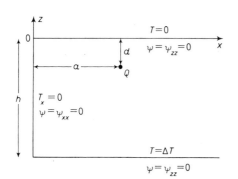

Fig. 1. – Geometry and boundary conditions for a drifting line source of heat.

everywhere the same. The Boussinesq descriptions of the dynamics and thermo-dynamics of this fluid are given in eqs. (1.1) and (1.2):

$$\text{(1.1)} \qquad \frac{D\omega}{Dt} + agT_x = \nu\nabla^2\omega \,,$$

$$\text{(1.2)} \qquad \frac{DT}{Dt} + \beta\psi_x = \varkappa\nabla^2 T + Q\delta(x-a)\,\delta(z+d) \,,$$

where $\beta = -\Delta T/h$; $D/Dt$ is the substantial derivative, subscripts indicate partial differentiation, and $\delta$ is the Dirac delta function. Here we require that $\Delta T$ always be less than the critical temperature difference needed to initiate free cellular convection. The relations of the stream function ($\psi$) to the vorticity ($\omega$), to the horizontal velocity ($u$), and to the vertical velocity ($w$) are:

$$\text{(1.3)} \qquad \nabla^2\psi = \omega \,, \qquad \psi_z = u \,, \qquad -\psi_x = w \,.$$

The problem is completed by eq. (1.4), which is the essential statement that the time derivative of source position is equal to the horizontal velocity in the fluid at the position of the source,

$$\text{(1.4)} \qquad a_t = \psi_z(a, -d) \equiv U \,.$$

A scaling of the variables of this problem based on the linear response ($V$) for fixed $a$ and $\beta = 0$, is:

$$\text{(1.5)} \qquad \left|\begin{array}{llll} \psi = (Vh)\,\psi' \,, & r = (h)\,r' \,, & T = \left(\dfrac{Q}{\varkappa}\right)T' \,, \\[2mm] \delta = (h^{-1})\,\delta' \,, & t = \left(\dfrac{h^2}{\varkappa}\right)t' \,, & V \equiv \left(\dfrac{agQh^2}{\varkappa\nu}\right) . \end{array}\right.$$

We now drop the primes in this scaling and rephrase the physics of our problem as

$$\text{(1.6)} \qquad \frac{1}{\sigma}\{\nabla^2\psi_t + R(\boldsymbol{v}\cdot\nabla)\nabla^2\psi\} + T_x = \nabla^4\psi \,,$$

$$\text{(1.7)} \qquad T_t + R\boldsymbol{v}\cdot\nabla T + R_a\psi_x = \nabla^2 T + \delta(x-a)\,\delta(z+d) \,,$$

$$\text{(1.8)} \qquad U = R\psi_z(a, -d) \,,$$

$$\text{(1.9)} \qquad \sigma \equiv \frac{\nu}{\varkappa} \,, \qquad R_a = \frac{ag\beta h^4}{\varkappa\nu} \,, \qquad R = \frac{Vh}{\varkappa} = \frac{agQh^3}{\nu\varkappa^2} \,,$$

in terms of the three independent physical parameters; the Prandtl number $\sigma$, the Rayleigh number $R_a$, and a thermal Reynolds number $R$.

It is relevant to the drifting continent problem to seek solutions of eqs. (1.6), (1.7), (1.8) for $1/\sigma$ vanishingly small. However, we would like to know the field of motion for $R$ both small and large. An expansion fo $\psi$ and $T$ in a power series in $R$ can give us valid solutions for small $R$ and some insight, perhaps, into the character of the solutions at larger $R$. The leading equations in an $R$ expansion, for vanishingly small $1/\sigma$, are

(1.10) $$T_x = \nabla^4 \psi \,,$$

(1.11) $$T_t + R_a \psi_x = \nabla^2 T + \delta(x - a)\,\delta(z - d) \,.$$

The next higher-order equations in $R$ are the linear inhomogeneous set

(1.12) $$T_{1x} = \nabla^4 \psi_1 \,,$$

(1.13) $$T_{1t} + R_a \psi_{1x} - \nabla^2 T_1 = - \boldsymbol{v} \cdot \nabla T \,,$$

whose solution depends on the solutions found for eqs. (1.10), (1.11).

To solve eqs. (1.10), (1.11), one first constructs the Fourier transforms appropriate to our boundary conditions,

(1.14)
$$
\begin{cases}
\theta(k,\,m) = \int\limits_{0}^{\infty} \cos{(kx)}\,\mathrm{d}x \int\limits_{-1}^{0} T(x,\,z)\,\sin{(m\pi z)}\,\mathrm{d}z \,,\\[2ex]
\varphi(k,\,m) = \int\limits_{0}^{\infty} \sin{(kx)}\,\mathrm{d}x \int\limits_{-1}^{0} \psi(x,\,z)\,\sin{(m\pi z)}\,\mathrm{d}z \,,
\end{cases}
$$

hence

(1.15)
$$
\begin{cases}
T(x,\,z) = \dfrac{4}{\pi} \int\limits_{0}^{\infty} \cos{(kx)}\,\mathrm{d}k \sum\limits_{m=1}^{\infty} -\theta(k,\,m)\,\sin{(m\pi)}(-z) \,,\\[3ex]
\psi(x,\,z) = \dfrac{4}{\pi} \int\limits_{0}^{\infty} \sin{(kx)}\,\mathrm{d}k \sum\limits_{m=1}^{\infty} -\varphi(k,\,m)\,\sin{(m\pi)}(-z) \,.
\end{cases}
$$

Then, from eqs. (1.10), (1.11), the equations satisfied by $\theta$ and $\varphi$ are:

(1.16) $$- k\theta = (k^2 + \pi^2 m^2)^2 \varphi \,,$$

(1.17) $$\theta_t + kR_a\varphi + (k^2 + \pi^2 m^2)\theta = - \cos{(ka)}\,\sin{(m\pi d)} \,.$$

The solution of eqs. (1.16), (1.17) for $\theta$ is, with $\theta = 0$ initially:

(1.18) $$\theta = - \sin{(m\pi d)} \exp{[-[\,]t]}\int\limits_{0}^{t} \exp{[+[\,]\tau]} \cos{(ka)}(\tau)\,\mathrm{d}\tau \,,$$

where

(1.19) 
$$[\ ] = \left[(k^2 + \pi^2 m^2) - \frac{R_a k^2}{(k^2 + \pi^2 m^2)^2}\right].$$

Hence, from eqs. (1.15), (1.16),

(1.20)
$$\left\{\begin{array}{l} \psi_z(a, -d) = \sum_{m=1}^{\infty} m \sin(2\pi d)\, m \int_0^{\infty} \frac{k\, dk}{(k^2 + \pi^2 m^2)^2} \int_0^t \exp\left[-[\ ]\eta\right]\{\ \}\, d\eta\,, \\ \eta \equiv t - \tau\,, \quad \{\ \} \equiv \{\sin k[a(t) + a(\tau)] + \sin k[a(t) - a(\tau)]\}\,. \end{array}\right.$$

With eq. (1.4), eq. (1.20) constitutes a formal solution to the first-order problem. However, eq. (1.20) also contains higher-order features of the flow because $a(t)$ has not been linearized. It is instructive to include the first advective effects of the source by considering the expansion:

(1.21)
$$a(\tau) = a(t) - a_t \eta + \tfrac{1}{2} a_{tt} \eta^2 + \ldots\,.$$

Then the retention of both $a(t)$ and $a_t \eta$ terms in eq. (1.21) for use in eq. (1.20) corresponds to the inclusion of both first and second-order $R$ terms for the source. Alternatively, one can restate the problem as that of determining the $R$ which will produce a linear flow at the source equal to an imposed constant source velocity $U$.

With the neglect of $a_{tt} \eta^2$ and higher terms in eq. (1.21), one can explicitly evaluate the time integral in eq. (1.20) as:

(1.22)
$$\left\{\begin{array}{l} \displaystyle\int_0^t \exp\left[-[\ ]\eta\right]\{\ \}\, d\eta = F \sin(2ak) + G[1 - \cos(2ak)]\,, \\[2mm] \displaystyle F = \frac{\{-[\ ]\cos(kUt) + kU \sin(kUt)\} \exp\left[-[\ ]t\right] + [\ ]}{[\ ]^2 + k^2 U^2}\,, \\[2mm] \displaystyle G = \frac{\{-[\ ]\sin(kUt) - kU \cos(kUt)\} \exp\left[-[\ ]t\right] + kU}{[\ ]^2 + k^2 U^2}\,. \end{array}\right.$$

For a certain time after the source $Q$ is first turned on the time-dependent terms in eq. (1.22) will be important. However, for

(1.23)
$$t \gg \frac{1}{[\ ]_{\min}} \equiv t_c$$

the time-dependence vanishes and

(1.24)
$$\int_0^{t \gg t_c} \exp\left[-[\ ]\eta\right]\{\ \}\, d\eta = \frac{[\ ]\sin(2ak) + kU[1 - \cos(2ak)]}{[\ ]^2 + k^2 U^2}\,.$$

Hence, the $k$ integral in eq. (1.21) may be written:

$$(1.25) \quad \int_0^\infty \frac{k\,\mathrm{d}k}{(k^2 + \pi^2 m^2)^2} \int_0^{t \gg t_0} \exp\left[-[\ ]\eta\right]\{\ \}\,\mathrm{d}\eta =$$

$$= \frac{i}{2} \int_{-\infty}^{+\infty} \frac{k(1 - \exp[i2ak])\,\mathrm{d}k}{\{(k^2 + \pi^2 m^2)^3 - k^2 R_a + ik U(k^2 + \pi^2 m^2)^2\}},$$

simplifying the form of eq. (1.24) by the use of complex notation. Then utilizing eqs (1.4) and (1.20), (1.25), one concludes that

$$(1.26) \quad U = \frac{R}{2\pi^4} \sum_{m=1}^\infty \frac{\sin(2\pi d)\,m}{m^3} \int_{-\infty}^{+\infty} \frac{k(1 - \exp[ibk])\,\mathrm{d}k}{(k^2 + 1)^3 - \gamma k^2 + ikv(k^2 + 1)^2},$$

where

$$b \equiv (2\pi a)\,m, \qquad \gamma \equiv R_a/\pi^4 m^4, \qquad v \equiv U/\pi m.$$

The problem remaining is to evaluate the $k$ integral and the $m$ sum as a function of $\gamma$, i.e. the Rayleigh number. This can be done readily for $R_a = 0$ and for $R_a$ close to its critical value for free convection. Intermediate values for $R_a$ involve the determination of the complex roots of the sextic equation

$$(1.27) \quad (k^2 + 1)^3 - \gamma k^2 + ikv(k^2 + 1)^2 = 0.$$

For $\gamma = 0$, the roots of eq. (1.27) are:

$$(1.28) \quad k = \pm i, \quad \pm i, \quad \frac{i}{2}\left(-v \pm \sqrt{v^2 + 4}\right).$$

Therefore

$$U = \left(\frac{R}{4\pi^3}\right) \sum_{m=1}^\infty \frac{\sin(2\pi d)\,m}{m^3} \left[\frac{v}{\sqrt{v^2 + 4}(2 + \sqrt{v^2 + 4})} + H \exp[-b]\right],$$

where

$$(1.29) \quad H \equiv \frac{2 + (1 + b)v}{v^2} - \frac{8 \exp\left[+b\left(1 - (\sqrt{v^2 + 4} - v)/2\right)\right]}{v^2 \sqrt{v^2 + 4}\left(\sqrt{v^2 + 4} - v\right)} \simeq$$

$$\simeq \frac{1}{4} b(1 + b), \qquad \text{for } v \ll 1.$$

It is seen that the $m$ series converges with great rapidity in most cases. For

$d = \frac{1}{4}$, roughly 0.98 of $U$ is contained in the first term, $m = 1$. Also from eq. (1.29), it is clear that the velocity of the initial separation of the two heat sources is proportional to $R$, dropping off exponentially with increasing separation. At distances large compared to $h/2\pi$, for $d = \frac{1}{4}$ and $m = 1$,

$$(1.30) \qquad U^2 = \pi^2 \left( \sqrt{\frac{R}{4\pi^4} + 1} + 1 \right) \left( \sqrt{\frac{R}{4\pi^4} + 1} - 3 \right).$$

Hence

$$(1.31) \qquad U \simeq \frac{1}{\sqrt{6}\,\pi} \sqrt{R - R_c}\,, \qquad\qquad R_c \equiv 32\pi^4$$

for

$$R \gg R_c$$

and

$$(1.32) \qquad U \simeq \frac{1}{2\pi} \sqrt{R - R_c}\,, \qquad\qquad R \gg R_c.$$

This latter conclusion, eq. (1.30), is entirely the consequence of retaining the second-order motion term, eq. (1.21). Hence the square-root dependence of $U$ on $R$ at great separation distance must be considered suspect until the remaining second-order terms in the flow field $\psi$ are determined.

The second case for which the $k$ integral can be determined without great labor is for

$$(1.33) \qquad \gamma = \frac{27}{4} - \varepsilon\,,$$

where both $\varepsilon$ and $v$ will be considered to be much smaller than one. We again choose $d = \frac{1}{4}$ and find that the $m = 1$ contribution to the flow is:

$$U \simeq \left( \frac{R}{2\pi^2} \right) \frac{2\sqrt{2}}{9} \frac{\pi}{p^2 + q^2} \left\{ q + \left( q \cos \left( \frac{1+v}{\sqrt{2}} \right) b - p \sin \left( \frac{1+v}{\sqrt{2}} \right) b \right) \exp - \frac{b\varepsilon}{q} \right\}$$

where

$$(1.34) \qquad \sqrt{\frac{2\varepsilon}{9} + i \frac{v}{\sqrt{2}}} = p + iq\,.$$

As in the $\gamma = 0$ case, one finds that $\dot{U}$ has an initial exponential decay, but here the decay is reduced by $\varepsilon/9$ and the velocity oscillates with a spatial period reflecting the free convection cells which would occur if $\varepsilon < 0$. At great separation distance, two different limiting velocities result from eq. (1.34). For an $\varepsilon$ chosen so that it is much smaller than $v$.

$$(1.35) \qquad U = 2\sqrt{2}\pi \left(\frac{R}{36\pi^4}\right)^{\frac{3}{2}}.$$

This result involves two limiting processes; ($\varepsilon \to 0$ and $R \to 0$) and suggests that one must reconsider the neglected time terms in eq. (1.22), for as $\varepsilon \to 0$, $t_c \to \infty$. However, for $\varepsilon \gg v$, one finds from eq. (1.34) that

$$(1.36) \qquad U = \frac{12\pi}{\sqrt{5}\sqrt{\varepsilon}} \sqrt{\left(\frac{R}{36\pi^4}\right)^2 - \frac{1}{4}\left(\frac{2}{9}\varepsilon\right)^3}.$$

These conclusions, eqs. (1.35), (1.36), indicate the significant increase in $U$ at a given $R$ which may occur as the stability of the fluid is decreased. However, these results are restricted to small values of $U$. A computation of eq. (1.26) for small $\varepsilon$ but large $U$ indicates that $U$ varies as the square root of $R$ as in the $\gamma = 0$ case.

## BIBLIOGRAPHY OF PART I

B. Isacks, J. Oliver and L. R. Sykes: *Journ. Geophys. Res.*, **73**, 5885 (1968).
C. L. Pekeris, *Month. Not. Roy. Astron. Soc., Geophys. Suppl.*, **3**, 343 (1935).

## PART II

## 2. – Self-convection of a block heat source.

This second, more geophysically oriented, model is pictured in Fig. 2. A two-dimensional block of width $L$ is immersed to a depth $(h - d)$ in a fluid

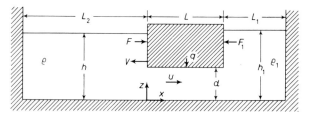

Fig. 2. – Geometry and boundary conditions for a drifting block source of heat.

of depth $h$. The block is a heat source of strength $q$ per unit length on its lower face. The fluid is contained in a region $L_1 + L_2 + L$ long with insulating side and bottom walls. To achieve a thermodynamically steady-state model we presume that the fluid is a uniform heat sink whose strength is sufficient to absorb the total heat flux, $qL$, from the block. This choice of heat sink is assumed to model a vertical heat flux from the upper surface of the fluid at the right and left of the floating block.

The principal assumptions of this model are: that $d$ is very small compared to $L$; that the flow, $u$, in the « channel » $d$ is laminar, *i.e.* that the channel's Reynolds number is small compared to one; that the flow in the regions to the right and left of the block cause density fluctuations which are small compared to $\varrho$-$\varrho_1$; that the thermal diffusion time across the channel, $d^2/\varkappa$, is small compared to the mass transport time along the channel $L/|u|$. The consistency of the following flow process is to be studied: starting with $L_1 \simeq 0$, the heating $qL$ causes the density of region $L_1$ to fall to $\varrho_1$; due to the resultant pressure head, the horizontal force on the block, $F_1$-$F$, gives rise to a block velocity, $V$, to the left; the resulting channel velocity, $u$, and heating, $qL$, are just sufficient so that the fluid emerging into region $L_1$ has decreased its density from $\varrho$ to $\varrho_1$. We assume that the heat capacity, $C$, the coefficient of thermal expansion, $\alpha$, and the kinematic viscosity, $\nu$, of the fluid are constants. It would

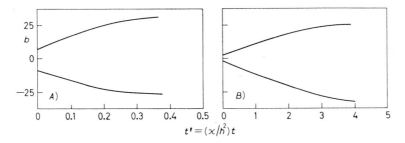

Fig. 3. – Time-distance trajectories for separating line heat source pairs at A) $R = 1.2 \cdot 10^3$ and B) $R = 5 \cdot 10^6$.

be compatible with this flow model to presume that the viscosity increased with height above the bottom, *i.e.* with temperature, so that the flow, $u$, was was confined to the bottom region until reaching the right hand wall. Ascending flow along the right hand wall then could be thought as of modelling the mid-Atlantic upwelling with symmetrically disposed second block moving off to the right of Fig. 3.

In order to use the simple Boussinesq description of the flow we will restrict attention to density contrasts, $\varrho$-$\varrho_1$, which are small compared to the density $\varrho$. Then, for a steady-state to be maintained

(2.1) $$F_1 - F = \frac{g}{2}\left[\varrho_1(h_1 - d)^2 - \varrho(h - d)^2\right] = -\nu\varrho Lu_z(d),$$

where $g$ is the acceleration of gravity, $u_z$ is the vertical gradient of $u$ at the lower boundary of the block, and the right-hand side of eq. (2.1) is the total viscous stress due to flow in the channel balancing the total horizontal pressure force, $(F_1 - F)$, on the vertical sides of the block. For thermal balance of the heat into and out of the channel one must have

(2.2) $$qL = \frac{Cd}{\alpha}(\bar{u} + V)\Delta\varrho,$$

where

$$\bar{u} \equiv \frac{1}{d}\int_0^d u(z)\,dz, \qquad \Delta\varrho \equiv \varrho - \varrho_1 = \varrho\alpha(T_1 - T).$$

Hence to determine $V$, one must compute the laminar Couette-Poiseuille-thermal flow, $u(z)$, in the channel. We seek solutions compatible with a pressure which varies linearly along the channel in the horizontal $(x)$ direction, so that

(2.3) $$P = g\left[\varrho(h - z)\left(1 - \frac{x}{L}\right) + \varrho_1(h_1 - z)\frac{x}{L}\right].$$

We are to find $u(z)$ from the horizontal equation of motion

(2.4) $$-\nu u_{zz} = -\frac{1}{\varrho}P_x$$

Subject to the boundary conditions

(2.5) $$u(0) = 0 \quad \text{and} \quad u(d) = -V.$$

The solution of eqs. (2.3)-(2.5) is:

(2.6) $$u(z) = \frac{gd^2}{2\varrho L\nu}\left(\frac{z}{d}\right)\left(1 - \frac{z}{d}\right)\left[\frac{\Delta P}{g} - \frac{\Delta\varrho}{3}(d + z)\right] - V\left(\frac{z}{d}\right),$$

where

$$\left(\frac{\Delta P}{g}\right) = \varrho h - \varrho_1 h_1.$$

Hence, from eq. (2.6), one finds that

$$(2.7) \qquad \bar{u} = \frac{gd^2}{12\varrho L\nu}\left(\frac{\Delta P}{g} - \frac{\Delta\varrho d}{2}\right) - \frac{V}{2}$$

and

$$(2.8) \qquad u_z(d) = -\frac{gd}{\varrho L\nu}\left(\frac{\Delta P}{2g} - \frac{1}{3}\Delta\varrho d\right) - \frac{V}{d}\,.$$

We note that

$$(2.9) \qquad \varrho_1 h_1^2 - \varrho h^2 = \varrho h^2\left(\frac{(1-\Delta P/\varrho gh)^2}{(1-\Delta\varrho/\varrho)} - 1\right) \simeq \varrho h^2\left[\frac{\Delta\varrho}{\varrho} - 2\frac{\Delta P}{\varrho gh}\right],$$

for $\Delta\varrho/\varrho \ll 1$ and $\Delta P/\varrho gh \ll 1$.

Therefore, with the use of eqs. (2.8) and (2.9), the force balance equation (2.1) may be written:

$$(2.10) \qquad \frac{1}{2}\,(h^2 - d^2)\Delta\varrho - (h-d)\frac{\Delta P}{g} = \frac{d}{2}\frac{\Delta P}{g} - \frac{d^2}{3}\Delta\varrho + \frac{\varrho L\nu}{gd}\,V\,.$$

Hence eqs. (2.2), (2.8) and (2.10) represent three equations for the four unknowns $\bar{u}$, $V$, $\Delta P$ and $\Delta\varrho$. A fourth equation relating these variables is found from the conditions for the continuity of mass flow between regions to the right and left of the drifting block. For the left region:

$$(2.11) \qquad L_2\frac{\partial\varrho h}{\partial t} = \varrho(h-d)V - \varrho\,d\bar{u}$$

and for the right region:

$$(2.12) \qquad L_1\frac{\partial\varrho_1 h_1}{\partial t} = -\varrho_1(h_1 - d) + \varrho\,d\bar{u}\,.$$

For a steady solution to eqs. (2.7), (2.8) and (2.10), both $\Delta P$ and $\Delta\varrho$ must be independent of time. From eqs. (2.11) (2.12) one finds that $\Delta P$ is independent of time when

$$(2.13) \qquad \varrho\,d\bar{u} = \left[\varrho(h-d) - \left(\frac{\Delta P}{g} - d\Delta\varrho\right)\left(1 - \frac{L_1}{L_1 + L_2}\right)\right]V \simeq \varrho(h-d)V\,,$$

the right-hand relation holding for small $\Delta\varrho$ and $\Delta P$. We will return to the conditions imposed on $\varrho$, $\varrho_1$, $h$, $h_1$ by the time-independence of $\Delta\varrho$ in the following Section. Equation (2.13) completes the set of equations needed to determine solutions for $\bar{u}$, $V$, $\Delta P$ and $\Delta\varrho$. Then from eqs. (2.2), (2.8), (2.10) and (2.13) we find that

$$(2.14) \qquad V^2 = \left(\frac{\alpha g q h^2}{24 \varrho C \nu}\right) F\left(\frac{d}{h}\right),$$

where

$$F\left(\frac{d}{h}\right) = \left(\frac{d}{h}\right)^3 \left[1 - \frac{d}{h} + \frac{1}{6}\left(\frac{d}{h}\right)^2\right] \Big/ \left[1 - \frac{d}{h} + \frac{1}{3}\left(\frac{d}{h}\right)^2\right].$$

One finds also that

$$(2.15) \qquad \Delta \varrho = \left(\frac{\alpha g L}{Ch}\right)\frac{1}{V}$$

and that

$$(2.16) \qquad \frac{\Delta P}{g} = \left(\frac{12 \varrho L \nu (h - d/2)}{g d^3}\right) V + \left(\frac{\alpha q L d}{2 Ch}\right)\frac{1}{V}.$$

The steady solutions eqs. (2.14)-(2.16) are possible when $\Delta \varrho$ as well as $\Delta P$ are time-independent. In addition, eqs. (2.11), (2.12) for mass continuity require that when $\Delta P$ is constant, then, independently

$$(2.17) \qquad \frac{\partial \varrho h}{\partial t} = 0, \qquad \frac{\partial \varrho_1 h_1}{\partial t} = 0$$

to first order in the small quantities $\Delta P$ and $\Delta \varrho$. A final relation determining the time-dependence of $\varrho$, $\varrho_1$, $h$, $h_1$ is the requirement that the total production of heat is balanced by a uniform sink of heat in the fluid. This latter condition, plus the constancy of $\Delta \varrho$ and eq. (2.17) leads to

$$(2.18) \qquad \frac{\partial \ln \varrho}{\partial t} = \frac{\partial \ln \varrho_1}{\partial t} = -\frac{\partial \ln h}{\partial t} = -\frac{\partial \ln h_1}{\partial t} = -\frac{\alpha q}{C \varrho^2 h}\left(\frac{L_1 + L_2}{L} + \frac{d}{h}\right)^{-1}.$$

We conclude that the drifting block moves from the right boundary plane to the left at a constant speed given by eq. (2.14), uniformly creating as much new fluid of density $\varrho_1$ as is destroyed by the uniform sink. Upon reaching the left boundary plane, the block reverses its direction and moves to the right. The turn-around time depends on the magnitude of the (neglected) second-order terms in $\Delta \varrho$ and the thermal diffusion velocity in the channel. Hence, $\Delta \varrho$ can always be chosen sufficiently small so that the turn-around time is a negligible fraction of the block transit time, $(L_1 + L_2)/V$.

The most significant conclusion is that the velocity $V$, eq. (2.14), depends on the fluid parameters and the heating rate in the same manner as does the final velocity $U$, eqs. (1.30)-(1.32) for the drifting line source treated in Part I.

The two expressions differ in that eq. (2.14) contains the geometric factors $(h/L)$ and $(d/h)$, and in that there is no value of heating or depth $d$ for which the motion vanishes.

## 3. – Laboratory experiments with drifting heat sources.

In this Section the first results from a continuing study of drifting heat sources is presented. The experimental arrangement simulates the line heat source model of Part I since that theory provides a description of the entire flow field.

The principal observable with which we were concerned was the velocity of the source as a function of the heating rate, geometric and fluid parameters. We sought also to discover whether an isolated heat source could propagate itself due to its own fluid motion, as is suggested in the results of Part I.

A variety of floats were used, all of which were heated by passing electric current through stainless steel wires. The electricity reached the floats through 0.005 cm diameter copper wires. When hung limply from above these lead wires produced negligible forces on the floats. The two types of sources reported upon here are: first, floating polyethylene strips with wires stretched beneath them; and second, wires contained in hollow aluminum oxide tubes floated by stryofoam pontoons at their ends.

Silicone oils of various viscosities were used as the fluid. The tank containing this fluid was 30 cm wide, 40 cm long and 10 cm high. The bottom of the tank was covered with $\frac{1}{2}$ cm of mercury in order to approximate the isothermal, slip boundary conditions used in the theory. The fluid was bounded above by air which, although providing the appropriate slip boundary condition, had two disadvantages. The first was that convection into the air did not provide a sufficiently isothermal surface. The second was that surface tension gradients, due to this variation of surface temperature, could be large enough to influence the source velocity. Although an experimental method exists to remove these difficulties, it was found that their effect on the observations could be adequately estimated.

Preliminary experiments with the polyethylene floats made us aware of the importance of any asymmetry in the heat source-float geometry. When a heating wire was positioned even slightly to one side of the center of its supporting floatation strip, invariably they would move in the direction of the less heated side. We concluded that only the very symmetric oxide tube source was suitable to simulate the theoretical line source of Part I. However, some data for an asymmetric source are included in Fig. 4 and will be discussed shortly.

The typical time trajectories of two oxide tube floats, for $R_x = 0$ and for two very different values of $R$, are shown in Fig. 3. The most striking feature of these observations is the constancy of the velocity of separation for many dimensionless length units after the power is first turned on. The second feature of importance is the gradual reduction in velocity at large separation.

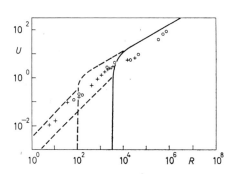

Fig. 4. – Velocity of separation, theory (solid line) and experiments (see text).

Neither result is in keeping with the theoretical deduction eq. (1.29). In the theory, the initial velocity of separation should depend linearly on $R$ and drop off as exp $[-b]$. The final velocity should be constant and depend linearly on the square root of $R$. We have concluded, tentatively, that: a thermal boundary layering process resulting from higher-order terms in the $R$ expansion is responsible for the initial constant velocity at moderately large $b$. We conclude, also, that the solution for great separation distance, eq. (1.30), is not stable for highly symmetric sources, at least not sufficiently stable to overcome the small three-dimensional end effects and float drag. However, the magnitude of the velocity of the source is in keeping with the theory, as is reported in the following paragraph.

A series of experiments were performed to determine the initial velocity as a function of $R$. The value of $R$ was changed by varying the depth of 600 stokes oil, the depth varying from 5 cm to 0.15 cm. Only two different heating powers were used, 0.5 W per cm and 1 W per cm. In this way it was hoped that spurious forces due to surface tension, and effects and float drag would be the same in all the experiments. The data for these experiments are plotted in Fig. 4 as $(+)$. The solid line in this Figure is the theoretical result eq. (1.32) in which $U$ varies as the square root of $R$. The lower dashed line represents the maximum value of $U$ $vs.$ $R$ from eq. (1.29). The upper dashed line is an approximate solution similar to eq. (1.29), but with the upper boundary of the fluid taken to be an insulator rather than isothermal. Evidently, there is some reasonable agreement with theory, which may be improved if float drag and end effects can be reduced.

The circular data points $(\circ)$ in Fig. 4 are for a single float which was purposely made asymmetric. It is seen that its velocity dependence on $R$ is very similar to that of the parting float pairs. Dr. NEWELL, in conversation, pointed out to us that source asymmetry could easily be incorporated into the analysis of Part I. If the line source in eq. (12) is taken as $[\delta(x-a) + \gamma\delta'(x-a)]$, where $\gamma \leqslant 1$ and $\delta'$ is the derivative of the delta function, then the source velocity becomes proportional to $\gamma$ at large separation distances.

The preceding data were taken with no heating from below, *i.e.* for $R_a = 0$. A very few experiments have been made for $R_a$ close to its critical value for free convection. A typical trajectory for $R_a = 0.9(R_a)_c$ is shown in Fig. 5. The fluid, floats and geometry were the same as those of Fig. 4. It is seen that the floats separate until they reach a distance $b = 12.5$, after which they maintain this separation and move off together at a reduced velocity. As one might have anticipated, this spacing is the exact size of the free convection cell which would first occur if the fluid were unstable $[R_a > (R_a)_c]$. It seems likely that the final velocity of the pair of floats is due to a slight difference in their source strengths. No studies have been made yet in the range $R_a > (R_a)_c$. It will be of considerable interest to discover the area in the $R, R_a$ plane in which the floating sources dominate the convective process.

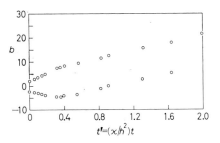

Fig. 5. – A drifting source with subcritical heating from below. $R = 2.7 \cdot 10^3$, $R_a = 0.9 \ (R_a)$ critical.

## 4. – Geophysical implications.

The experiments and the theories of both Parts I and II suggest that source drift velocities proportional to the square root of the heating rate can be expected to occur in a fluid differentially heated by the floating sources. Theory and experiment also suggest that the drift velocity is independent of the thermometric conductivity of a fluid with a large Prandtl number.

A qualitative consequence of the block model is that the heat flux $q$ emerges from the « ocean » regions, although it is produced in the « continent ». We presume, of course, that a similar heat flux emerges from the top of a realistic continent. In applying this model to the Earth one must add the horizontally homogeneous heat sources and the appropriate adiabatic lapse rate to the Boussinesq description of the vertical temperature distribution. Hence the model does not require that heat actually flow down, but only that the differential heating produces temperature variations in the horizontal.

To determine the magnitude of drift velocity from eq. (2.14), we have assumed that the differential heating rate between continent and ocean is approximately $q = 10^{-6}$ cal/s cm. We have chosen $h = 6.10^7$ cm, as this is the largest depth at which there is evidence for significant motion. The source depth is taken to be $(h - d) = 3 \cdot 10^6$ cm. However $F(d/h)$ of eq. (2.14) is quite insensitive to this choice. The principal uncertainty is the value of the kinematic viscosity. PEKERIS (1935), and TURCOTTE and OXBURGH, (1969 $a$, $b$), have used Haskell's (1935) value of $\nu = 3.10^{21}$ poise, estimated from the Fennoscandian post-glacial uplift. The parameter group ($\alpha g / \varrho C$) is taken to be $2.5 \cdot 10^{-2}$ cm$^4$/cal s$^2$ with confidence that the uncertainty in this value is small compared to the uncertainty for the values of $\nu$ and $h$. With these estimates used in eq. (2.14) we find that

$$V_{\text{continent}} \simeq 3.10^{-8} \text{ cm/s} = 1 \text{ cm/yr} .$$

As a consequence, the density contrast $\Delta \varrho / \varrho$ across a continent of $L = 5 \cdot 10^8$ cm is less than $10^{-2}$ from eq. (2.15). Also, one finds from eq. (2.16) that the difference in pressure head across the continent is approximately $3.10^5$ cm. A discussion of the strength of the crustal material and whether the estimated forces are sufficient to account for the observed breaking, folding, and flowing will not be attempted here.

## BIBLIOGRAPHY OF PART II

N. A. HASKELL: *Physics*, **6**, 265 (1935).

C. L. PEKERIS: *Month. Not. Roy. Astron. Soc., Geophys. Suppl.*, **3**, 343 (1935).

D. L. TURCOTTE and E. R. OXBURGH: *Journ. Geophys. Res.*, **74**, 1458 (1969).

D. L. TURCOTTE and E. R. OXBURGH: *Physics Today* (April 1969).

# Influence of the Core on the Nutations.

R. O. Vicente

*Faculdade de Ciencias - Lisbon*

We are interested in the study of the nutations but mainly concerned about the influence of the core on these motions. We have therefore to consider the following cases: 1st) the Earth is a rigid and homogeneous solid—this hypothesis has long been employed in astronomy; 2nd) the Earth is an elastic and heterogeneous body—this has been considered in geophysics.

The dynamical equations of the motion of a rigid body can be expressed by the vector equation

$$(1) \qquad \frac{d\boldsymbol{H}}{dt} = \boldsymbol{G} ,$$

showing that the time derivative of the angular moment $\boldsymbol{H}$ around the centre of mass is equal to the vector $\boldsymbol{G}$ of the exterior forces. The projection of this equation on the system $Oxyz$, coinciding with the principal axes of inertia of the Earth, corresponds to the well-known Euler equations:

$$(2) \qquad \begin{cases} A\dot{\omega}_1 + (C - B)\omega_2\omega_3 = L , \\ B\dot{\omega}_2 + (A - C)\omega_3\omega_1 = M , \\ C\dot{\omega}_3 + (B - A)\omega_1\omega_2 = N , \end{cases}$$

where $(A, B, C)$ are the principal moments of inertia, $(L, M, N)$ the moments of the external forces and $(\omega_1, \omega_2, \omega_3)$ the components of the rotation $\boldsymbol{\omega}$. The motion of the Earth in relation to the centre of mass $O$ can be represented, in any instant, by the instantaneous rotation $\boldsymbol{\omega}$ on an axis called the instantaneous axis of rotation or simply the axis of rotation.

These equations show that to determine the motion of the Earth, within given conditions, the moments of inertia are the only parameters that have a dynamical meaning and so it is not necessary to consider the geometrical form of the Earth.

The external forces are due to the Sun and Moon acting on the solid body of the Earth. As is well-known, the theories of the Sun and Moon present some inaccuracies which limit the final precision obtained for the theory of the motion of the Earth around its centre of mass.

The systems of co-ordinates employed for the theory of the motion of the Earth show the importance of the following axes:

1) the axis $Oz$, called the axis of figure, about which the moment of inertia is maximum;

2) the fixed axis $OZ$ directed towards the north pole of the ecliptic;

3) the axis of rotation, along which is the instantaneous rotation $\boldsymbol{\omega}$ that represents the motion of the Earth at every instant;

4) the axis along the vector angular momentum $\boldsymbol{H}$, through the centre of mass, and called the invariable axis in the dynamics of rotating rigid bodies.

The Earth has several motions around these axes with periods ranging from at least a day till several tens of thousands of years. It is hard to forecast the influence of so different periods on the behaviour of the Earth subject to such short and long time intervals.

The actual motion of the Earth can be considered as a steady state of rotation perturbed by external forces. This way of looking at the problem is justified because the effects of the external forces, due to the Sun and Moon, are comparatively small.

To find this steady state we consider that the external forces do not exist $(\boldsymbol{G} = 0)$, and the equations take the form:

$$(3) \quad \begin{cases} A\dot{\omega}_1 + (C-B)\,\omega_2\omega_3 = 0 \,, \\ B\dot{\omega}_2 + (A-C)\,\omega_3\omega_1 = 0 \,, \\ C\dot{\omega}_3 + (B-A)\,\omega_1\omega_2 = 0 \,. \end{cases}$$

One simplification which is made in the equations of motion is the hypothesis that the principal moments of inertia $A$ and $B$ are equal. Therefore the third equation shows immediately that

$$\omega_3 = \text{const} = n$$

and this result is valid in a first approximation, and we have detected irregularities in the daily rotation of the Earth only during the last 30 years.

The other two equations can be written in the following way:

$$(4) \quad \begin{cases} \dot{\omega}_1 + \dfrac{C-A}{A}\,n\omega_2 = 0 \,, \\ \dot{\omega}_2 - \dfrac{C-A}{A}\,n\omega_1 = 0 \,, \end{cases} \quad \text{or} \quad \begin{cases} \omega_1 = \sigma n \cos\,(pt+\beta)\,, \\ \omega_2 = \sigma n \sin\,(pt+\beta)\,, \end{cases}$$

where $\sigma n$ and $\beta$ are constants of integration, and $p = n(C-A)/A$.

The period of this motion, called the free period, is

(5)
$$\tau = \frac{2\pi}{p} = \frac{2\pi}{n} \cdot \frac{A}{C-A}.$$

Considering the sidereal day as the unit of time then $n = 2\pi$ and $\tau = A/(C-A)$. This ratio $A/(C-A)$ has been determined from the period of precession of the equinoxes; the value obtained for $\tau$ is about 305 sidereal days.

Referred to the set of axes $Oxyz$, coinciding with the principal axes of inertia of the Earth, the equation of the instantaneous axis of rotation is:

$$\frac{x}{\omega_1} = \frac{y}{\omega_2} = \frac{z}{\omega_3}$$

and substituting by the values obtained above, we have

(6)
$$\frac{x}{\sigma \cos (pt + \beta)} = \frac{y}{\sigma \sin (pt + \beta)} = \frac{z}{1}.$$

This expression shows that the instantaneous axis describes a circular cone about the axis of figure in the period $\tau$ (about 10 months). This motion is called the free Eulerian nutation; as an astronomical consequence the latitude of any place should exhibit a variation with a period of about 10 months. This phenomenon has been surveyed for nearly a century.

When there are no external forces, eq. (1) shows that the angular momentum is constant and the vector $H$ is fixed in space. The axis of the instantaneous rotation $\omega$, the axis of the angular momentum $H$ and the axis of figure all pass through the centre of mass of the Earth. Considering the components of the vectors $\omega$ ($\omega_1$, $\omega_2$, $\omega_3$), $H$ ($A\omega_1$, $A\omega_2$, $C\omega_3$) and the unit vector of the axis of figure, on the system of rectangular co-ordinates $Oxyz$, we have

(7)
$$\begin{vmatrix} \omega_1 & \omega_2 & \omega_3 \\ A\omega_1 & A\omega_2 & C\omega_3 \\ 0 & 0 & 1 \end{vmatrix} = 0 ,$$

that is, the axis of figure, the axis of rotation and the axis of angular momentum lie in the same plane. As the vector $H$ is fixed in space, and the axis of rotation describes a circular cone about the axis of figure, the plane rotates in space around $H$. The axis of figure and the axis of rotation lie on opposite sides of the vector $H$, describing circular cones in space.

We can determine theoretically the values of the angles between these 3 axes. The values obtained are:

   1) about $0''.3$ or 10 m at the surface of the Earth, for the angle $\gamma$ between the axes of figure and rotation, that is, the geographical poles are 10 m away from the poles of figure. This value shows that the component $\omega_3$ is nearly the whole of $\omega$.

   2) about $0''.001$ or 3 cm at the surface of the Earth, for the angle $\delta$ between the axis of rotation and the axis of the vector $\boldsymbol{H}$. This value means that $\boldsymbol{\omega}$ coincides with $\boldsymbol{H}$ from the point of view of astronomical observations, because we have not yet instruments that can determine the values of so small angles.

   We have now to consider the existence of external forces due to the Sun and Moon, therefore $\boldsymbol{G} \neq 0$ and the vector $\boldsymbol{H}$ is not fixed in space.

   This case is better studied considering the equations expressed in terms of the Euler angles ($\theta$ obliquity and $\psi$ longitude) and the equations obtained, considering the first terms only, are the well-known Poisson's equations:

(8)
$$
\begin{cases}
\dfrac{d\theta}{dt} = -\dfrac{1}{Cn \sin \theta} \dfrac{\partial U}{\partial \psi}, \\[2ex]
\dfrac{d\psi}{dt} = \dfrac{1}{Cn \sin \theta} \dfrac{\partial U}{\partial \theta},
\end{cases}
$$

where $U$ is the gravitational potential of the external forces.

   The more recent research on this subject has been done by WOOLARD [1]. For our purposes it is sufficient to know that the integration of this system of differential equations can be represented in the following way:

$$
\theta = \theta_p + \Theta, \qquad \psi = \psi_p + \Psi.
$$

   The part $(\theta_p, \psi_p)$ corresponds to the so-called *luni-solar precession* with terms of the type $at + bt^2 + ct^3 + \dots$ ($a, b, c, \dots$, constants) that increase continuously with time. The part $(\Theta, \Psi)$ corresponds to the *luni-solar nutation* with terms of the form

$$
a_1 \sin(\alpha_1 t + \beta_1) + a_2 \sin(\alpha_2 t + \beta_2) + \dots,
$$

where $a_1, a_2, \dots, \alpha_1, \alpha_2, \dots, \beta_1, \beta_2, \dots$, are suitable constants; these terms are periodic functions of time.

   The theory employed at present considers up to 70 terms in these expansion in series. The more important terms are: the lunar nutation with a period of 18.6 years and an amplitude in obliquity $9''.21$ being one of the fundamental constants in astronomy called the constant of nutation;

the semi-annual solar nutation with an amplitude of 0″.55 in obliquity; the fortnightly lunar nutation with an amplitude of only 0″.088 in obliquity; the luni-solar precession being the smooth long-period motion of the mean pole of the equator round the pole of the ecliptic, with a period of about 26 000 years and a value in longitude 50″.2, called the constant of precession, also one of the fundamental constants of astronomy.

This theory has been constructed considering 4 simplified hypotheses: 1) the Earth is a rigid body; 2) the principal moments of inertia $A = B$; 3) the Earth is considered as homogeneous; 4) the influence of the oceans and the irregular distribution of the continents are not considered.

From the geophysical point of view, the motions with probably greater influence are: the daily rotation with a period of one sidereal day, the free Eulerian nutation with a period of nearly 10 months, the lunar nutation with a period of about 19 years and the luni-solar precession with a period of nearly 26 000 years. All these motions, deduced theoretically from the simple theory of the motion of a solid around its centre of mass, have a permanent effect on any other temporary or permanent motion of the Earth.

We have now to compare the theoretical values with the observed values, computed from astronomical observations:

|  |  | Theory | Observation |
|---|---|---|---|
| forced nutations | lunar nutation (period 18.6 years)<br>semi-annual solar nutation<br>fortnightly lunar nutation | 9″.226<br>0″.5528<br>0″.0945 | (9′.198 ±0″.002)<br>(0″.578 ±0″.004)<br>(0″.0949±0″.0010) |
|  | free Eulerian nutation | 10 months | 14 months |

It took quite a time to realize that the disagreements were due to the fact that the Earth did not behave as simply as the theoretical model. NEW-COMB [2] explained the disagreement between the theoretical and the observed values of the variation of latitude saying that the observations showed the elastic behaviour of the Earth, therefore the first hypothesis was abandoned; so the fact that the Earth is not a rigid body was obtained exclusively by astronomical observations.

The disagreements that appear in the forced nutations could only be explained after the seismological studies permitted the determination of the density and elastic parameters of the Earth's interior, therefore the third hypothesis had to be abandoned.

The hypothesis that the moments of inertia $A$ and $B$ are equal has no influence in our results; the difference, as it is presently known, is not significant for the precision of the observations made nowadays.

During the second half of the last century there were a number of researches
in mathematical physics concerning the problem of the motions of a mass
of fluid contained in a rigid envelope. For instance, KELVIN [3] pointed out
that if a portion of the Earth were fluid the theoretical value of the constant
of nutation would be reduced. The researches of Hough [4] concerning the
oscillations of a rotating ellipsoidal mass of liquid contained in a rigid envelope
showed that the period of the Eulerian nutation would be shortened if the
Earth's core were considered as fluid. Hough's investigation is very detailed.
He takes into consideration the ellipticity of the Earth. The problem is solved
by elliptic functions.

In those days there were no indications about the possibility that the central
part of the Earth were liquid; the values obtained differed more from the observational values than when a rigid model was adopted, therefore this line of
investigations was abandoned. If it had been pursued, considering the elasticity of the shell, we could say that the existence of a liquid core for the Earth
could have been forecast by the theory at the end of last century instead
of having to wait for the experimental results of seismology.

Considering several models for the Earth (VICENTE [5]), we can obtain the
following values:

| Model | Free Eulerian nutation (sidereal days) |
|-------|----------------------------------------|
| Rigid solid | 305 |
| Elastic solid | 319 |
| Rigid shell, liquid core | 272 |
| Elastic shell, liquid core | 401 |

that justify the preceding remarks.

The consideration of an elastic Earth with a liquid core gives the greatest
value for the period, but they are not sufficient to obtain agreement with the
observed 14 months period. It is necessary to consider the influence of the
oceans and the obstruction caused by the irregular distribution of the continents, so the remaining simplified hypothesis has to be abandoned; we cannot
yet compute this influence with enough accuracy, but the values determined
give an order of magnitude of 30 to 40 days, bringing into agreement the theory
and the observations.

Once it was realized that the Earth behaved as an elastic body one began
to consider another aspect of the actions of the Sun and Moon on the Earth:
the tidal attractions of the luni-solar forces, giving rise to the tides of the lithosphere. The nutations and the bodily tides have been studied independently

of each other till recently; originating in the same forces, the theory trying to explain them should be a unified theory.

Different components of the Earth tides produce the nutations that we have considered; the forced nutations correspond to diurnal tides.

The problem of the Earth tides was first considered by KELVIN [6] who discussed the elastic deformation of a homogeneous incompressible sphere of the size of the Earth. Many researches have been made after the work of Kelvin, considering several Earth models with different values for the elastic parameters and the distribution of density, but only recently it has been possible to get reliable results about the interior of the Earth.

All the studies concerned with a statical theory applied to an elastic solid Earth with spherical symmetry introduce the numbers $h$, $k$ and $l$ that can be defined in the following way (JEFFREYS [7]): If the disturbing potential $U_2$ is a solid harmonic of second degree we can define an equilibrium tide as equal to $U_2/g$, where $g$ is the undisturbed gravity at the surface $r = a$. We consider the radial displacement at the surface to be $hU_2/g$ and the deformation of the Earth to produce a disturbance of the gravitational potential whose value at the surface is equal to $kU_2$. The horizontal displacements of a particle at the surface $r = a$ may be written:

$$u_\varphi = \frac{l}{g}\frac{\partial U_2}{\partial \varphi}, \qquad u_\lambda = \frac{l}{g\cos\varphi}\frac{\partial U_2}{\partial \lambda}$$

($\varphi$ geographic latitude and $\lambda$ longitude), giving the definition of $l$.

The value of $k$ is related to the period of the Eulerian nutation and the value of $(1 + k - l)$ to the disturbances that affect the position of the astronomical vertical of a place on the Earth (variations in the geographic co-ordinates) with periods depending on the periods of the tidal forces. In practice, the determination of the values of $k$ and $(1 + k - l)$ is made more difficult because there are local corrections. The values of $h$, $k$ and $l$ can also be determined from geophysical observations and, more recently, from satellite observations.

Most researches done on the bodily tides express their results in terms of these numbers, with a great advantage: whatever theoretical Earth model or instrumental technique is adopted, we always have the same standard of comparison for expressing the results.

The solution of the problem of the bodily tide of the Earth, adopting an Earth model in fairly good agreement with the observations, was first done by TAKEUCHI [8] by numerical methods, but not yet employing electronic computers. The problem is reduced to the solution of a system of 3 differential equations of second order. The bodily tide numbers corresponding to Earth models based on Bullen's values of the density and elastic parameters agree fairly well with the values known from the observations.

After the choice of the Earth model has been made, it is necessary to investigate whether a dynamical or a statical theory is convenient for both the shell and the core, or only for one of them. The nutations are connected with the diurnal tides, but there are also semi-diurnal, fortnightly and semi-annual tides among the more important. Since the earlier investigations have shown that the periods of free oscillations of tidal type are of the order of one hour for any admissible constitution of the Earth, we can consider these periods as short in comparison with the diurnal tides. Therefore it will be sufficient to employ a statical theory at least for the shell of the Earth model considered. This is already an important result.

We have now to examine what happens for the core: JEFFREYS [9] has shown that the application of a statical theory is still valid for the semi-diurnal, forthnightly and semi-annual tides because the errors introduced (neglecting rigidity and inertia) are of the same order of magnitude as the approximations made considering the Earth as spherical. For the diurnal tides the neglect of rigidity and inertia is no more valid because of the boundary conditions at the core.

Another fact we must have in mind is that the free and forced nutations considered tend to alter the position of the axis of rotation of the Earth, and this means that the boundary conditions at the core cannot be satisfied by a statistical theory with the precision necessary for the solution of the problem. Conversely there is an influence of the nutations on the motions of the core.

The actual theory of the nutations and bodily tides proposed by JEFFREYS and VICENTE [10] takes in consideration these facts, adopting an Earth model composed of a shell and a core in such a way that it can use the results obtained by TAKEUCHI for the shell; for the core it was considered sufficient to adopt two models corresponding to extreme cases of its possible behaviour. The core is always considered as liquid, in the sense that $\mu \approx 0$.

The central-particle model considers the core as homogeneous and incompressible, with a point mass added at the centre to give an estimate of the possible effects of the inner core. The expressions of the displacements are linear functions of the co-ordinates.

The Roche model considers a density distribution in the core that corresponds to a Roche type of density distribution law. In this model the displacements are not linear functions of the co-ordinates and their expressions are such that they are intended to represent the main motions in the core.

The values of the bodily tide numbers computed for the statical case are:

|                          |   $h$   |   $k$   |   $l$   |
|--------------------------|---------|---------|---------|
| Central-particle model   |  0.585  |  0.289  |  0.082  |
| Roche model              |  0.598  |  0.273  |  0.082  |

The values obtained are practically the same because we are considering shells with the same composition; the mass and moment of inertia adopted for the core are also the same. We can conclude that any suitable model of the core will not affect the values of $h$, $k$ and $l$ within the hypotheses adopted in this theory.

The solution of the equations of motion for the forced nutations leads to the determination of the ratio $\zeta/\zeta_0$ where $\zeta$ represents the motion corresponding to the Earth model adopted and $\zeta_0$ is the value of $\zeta$ for a rigid Earth taken as a standard of comparison.

Considering the case of the lunar nutation with a period of 18.6 years, the values adopted by NEWCOMB are $9''.210$ from the observations and $9''.220$ from the theory, so we get the value of 0.9989 for the ratio. The more recent determinations give 0.996 or 0.997, depending on the observed values adopted for the constant of nutation. The ratio $\zeta/\zeta_0$ for the central-particle model is 0.996 and for the Roche model is 0.999, that is, the models considered do represent the actual behaviour of the Earth.

The other nutations with smaller values, like the fortnightly and semi-annual, have been determined from the observations only during the last 15 years, because of the difficulties in the astronomical observations which correspond to the limit of precision obtainable at present. The comparison with the theory (JEFFREYS [11]) shows that the semi-annual nutation agrees best with the central-particle model; the forthnightly nutation is in agreement with both models. The detailed analysis of the results shows that the relationship between the core and the nutations is a complex one.

Another theory of the bodily tide and nutations was developed later by MOLODENSKY [12, 13] considering several Earth models. One model considers the variation of density in the core as due to compression alone and the other model considers the existence of an inner core. A good explanation of Molodensky's theory is given by JOBERT [14]. The values obtained for the bodily tide numbers are in fairly good agreement with the values indicated above. A comparison of the values obtained from different forms of the elastic equations for the Earth has been made by JEFFREYS and VICENTE [15].

This outline of the theory, and the comparison with the observed values, gives us an indication about the difficulties encountered:

*a*) The great difficulties in obtaining reliable observed values, specially for the nutations of smaller amplitudes (from the fact that the theory is in good agreement for the nutations of greater amplitude, we can infer that the values predicted by the theory will give an indication of the order of magnitude for the smaller ones).

*b*) The difficulties in choosing convenient Earth models.

The lack of geophysical observations giving us a detailed knowledge about the constitution and properties of the Earth's interior has been largely removed during the last decades, the boundary layer of the core remaining one of the most important sources of uncertainty.

The difficulties of solving, by numerical methods, the systems of differential equations representing the motions of the Earth have disappeared with the invention of electronic computers, which have led conversely to a proliferation of Earth models, creating another difficulty—what models should we choose? To have not yet adopted a standard model for the Earth, as it has been done for instance in geodesy and in the researches on the bodily tide of the Earth, has impaired a critical discussion of many published results.

## REFERENCES

[1] E. W. Woolard: *Astr. Pap. Amer. Ephem. Wash.* **15**, part I (1953).

[2] S. Newcomb: *Mon. Not. Roy. Astr. Soc.*, **52**, 336 (1892).

[3] Lord Kelvin: *Math. And Phys. Papers*, **3**, 320 (1876).

[4] S. S. Hough: *Phil. Trans*, A, **186**, 469 (1895).

[5] R. O. Vicente: *Marées Terrestres, Bull. Inf.*, **53**, 2489 (1969).

[6] Lord Kelvin: *Phil. Trans.*, A, **153**, 573 (1863).

[7] H. Jeffreys: *The Earth*, 5th ed. (Cambridge, 1970).

[8] H. Takeuchi: *Trans. Amer. Geophys. Un.*, **31**, 651 (1950).

[9] H. Jeffreys: *Mon. Not. Roy. Astr. Soc.*, **109**, 670 (1949).

[10] H. Jeffreys and R. O. Vicente: *Mon. Not. Roy. Astr. Soc.*, **117**, 142 (1957).

[11] H. Jeffreys: *Mon. Not. Roy. Astr. Soc.*, **119**, 75 (1959).

[12] M. S. Molodensky: *Comm. Obs. Roy. Belgique*, **188**, 25 (1961).

[13] P. J. Melchior: *The Earth Tides* (London, 1966).

[14] G. Jobert: *Comm. Obs. Roy. Belgique*, **236**, 64 (1964).

[15] H. Jeffreys and R. O. Vicente: *Mem. Cl. Sci., Acad. Roy. Belgique*, II série, **37**, fasc. 3 (1966).

# Knowledge of the Earth's Core from Geomagnetism.

F. J. Lowes

*School of Physics, University of Newcastle upon Tyne - Newcastle upon Tyne*

## 1. – In roduc ion.

We now have a reasonable picture of the present geomagnetic field, and an increasing knowledge of its behaviour over the last $10^9$ years. (We are also starting to obtain information about the present magnetic fields of other planets.) In the last five years there has been a renewed interest in various aspects of the « dynamo » problem—the production and maintenance of the field—and we are beginning to understand rather more of what is involved.

It seems reasonable therefore to ask in what way geomagnetic observation and theory can contribute to our knowledge of the properties of, and processes in, the core of the Earth (and possibly of other planets). This paper attempts to answer this question as far as is possible at present.

In Sect. **2** the relevant observations are summarized, and in Sect. **3** a very brief outline is given of present thinking about the origin of the field. Sections **4, 5** and **6** then discuss possible deductions from the existence of the field (and some of its properties), Sect. **7** from its magnitude, and Sect. **8** from its reversals.

## 2. – The geomagnetic field.

As observed at the surface the Earth's magnetic field shows time variations with periods from about $10^6$ years to $10^{-3}$ s. Most of the field is due to electric currents in the conducting fluid core, some comes from the permanent and induced magnetization of crystal rocks, and the rest originates in the ionosphere and magnetosphere. Fortunately, practically all the fields of external origin have periods shorter than one year, so properly filtered data give the field of internal origin. Also, the magnetic « anomalies » due to crystal rocks mostly have length scales of less than about 1000 km, while the field coming

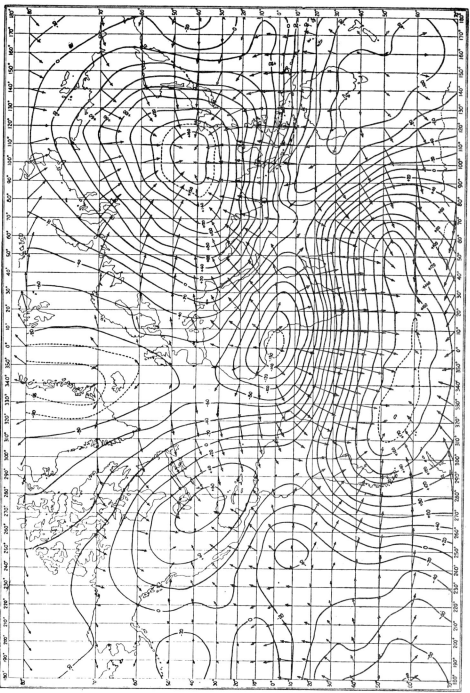

Fig. 1. – Nondipole field for 1945. The contours give the vertical component in units of $10^2 \gamma$. The arrows give the horizontal field; a length corresponding to $10°$ of longitude represents $6.6 \cdot 10^3 \gamma$. (Reproduced by permission. From BULLARD, FREEDMAN, GELLMAN and NIXON: *Phil. Trans. Roy. Soc.*, A **243**, 67 (1950).)

from the core, 3000 km deep, would be expected to have length scales greater than 1000 km. It is therefore possible to obtain a reasonable picture of the main magnetic field, that of deep internal origin.

Direct observations of the vector field over much of the Earth's surface are available for about 150 years, with more scattered, less precise, partial, observations going back for about another 100 years. At present the field is of magnitude $(30\,000 \div 60\,000)\,\gamma$ $(1\,\gamma = 10^{-5}\,G = 10^{-9}\,\text{Tesla})$ and is dominantly dipolar, the best-fitting dipole being close to the geocentre and inclined at about $10°$ to the geographic axis.

The difference between the best central dipole and the actual field is called the nondipole field (Fig. 1). It has r.m.s. magnitude of about $8000\,\gamma$ with maxima up to $17\,000\,\gamma$, and a length scale of some thousands of kilometers.

The time variation of the field of internal origin is called the secular variation (because of its comparative slowness on a human time scale). It has r.m.s. magnitude about $70\,\gamma/\text{year}$, with maxima up to $170\,\gamma/\text{year}$. The dipole field varies only slowly, up to about $20\,\gamma/\text{year}$ $(0.05\%/\text{year})$, and the major part of the secular variation comes from changes in the nondipole field, which is considerably altered in 100 years. As well as changing its pattern the nondipole field drifts westward at about $0.2°/\text{year}$; these two effects contribute about equally to the secular variation.

Data from further back come from the direction and sometimes intensity of magnetization of baked clays (archeomagnetism) and of some volcanic and sedimentary rocks (palaeomagnetism). Palaeomagnetic studies have shown that the Earth has possessed a magnetic field for at least $3 \cdot 10^9$ years. Also, the direction of the field shows reversals, the time between reversals being typically $10^5$ to $10^6$ years, and the duration of the reversals being $10^4$ years or less; these reversals have been observed back to $10^9$ years so far.

Data for nearer epochs suggest that the Earth's magnetic field has in many respects been similar to that at present. A reasonably good approximation to the field has been that of a dipole of either direction, with moment varying in the range $(0.3$ to $1.5)$ times the present, and with axis within about $10°$ of the Earth's spin axis; on average the dipole axis has been very close to the spin axis. The smaller scale data that exist are consistent with the departure from this dipole (the nondipole field) and the time variation of the field being roughly as at present.

## 3. – The origin of the field.

It is now generally agreed that the main field is due to electric currents in the conducting liquid core of the Earth. These currents are believed to be maintained by a «dynamo» action, motions of the fluid conductor through the magnetic field giving induced electromotive forces which maintain the currents against ohmic dissipation—a self-exciting dynamo.

Such a situation is easily understood in terms of the dynamos we use to generate our electricity. These generate current when the wires of the rotor are moved in the magnetic field produced by the stator windings; we simply use part of the output current to « excite » the stator. However such machines are multiply connected (in the topological sense) and are inhomogeneous (we rely on insulation round the wires to make the current flow in the right places).

For a long time the possibility of a self-exciting dynamo in a homogeneous simply connected fluid conducting sphere such as the Earth's core was disputed. However GOLD [1] showed that multiple connection was not important, and eventually BACKUS [2] and HERZENBERG [3] independently provided theoretical existence proofs; later LOWES and WILKINSON [4] produced a working homogeneous self-exciting dynamo in the laboratory.

Until recently the most plausible self-exciting dynamo process, and certainly the one most investigated, was the type introduced by BULLARD and GELLMAN [5]. In this two motions are involved—a laminar convective motion and a differential rotation—and are both of length scale comparable with that of the core. Recently however it has been shown that many spatially periodic motions of length scale small compared with that of the core could also maintain large-scale magnetic fields [6]. Similarly, some small scale turbulent motions having mirror asymmetry can also maintain large-scale magnetic fields [7].

We do not know if the nondipole field, is a direct consequence of the dynamo process, or whether it is an unconnected perturbation. There is some evidence that during a reversal, while the « dipole » field goes through « zero » a field of the magnitude of the nondipole field remains.

In view of the great uncertainties of the subject I will try to restrict my deductions to those which are not critically dependent on particular dynamo models.

## 4. – Restrictions on the structure of the Earth (or planet).

Even before dynamo maintenance was accepted there were very strong indications that the field must originate in a fluid. The time scale of the variations of the nondipole field is 100 years or less, and this variation is so much faster than any other « solid Earth » phenomenon that it is inconceivable that it could arise in solid materials. The very pattern of the nondipole field is also highly reminiscent of a meteorological chart.

Quite clearly for a planet to produce a magnetic field, there must be a region which is fluid and electrically conducting. We now know this is so in the case of the Earth, but it is interesting that, in the very first suggestion of dynamo action in 1919, LARMOR [8], who was primarily discussing the sun, suggested that it would also work for the Earth if there were such a core; this was at the time when seismology was « producing » the core.

However the fluid conductor need not necessarily be a central core; HIDE [9] has suggested that in Jupiter it might be a shell near the outside.

I will discuss possible restrictions on the size of the conducting region later.

## 5. – Restrictions on the motions.

The fact that the geomagnetic dipole is very close to the geographic axis is surely significant, so the fact that the Earth rotates must be important. Certainly, order-of-magnitude calculations show clearly that the Coriolis and Lorentz forces are dominant in the Earth's core. In the small scale motion dynamos there has to be some mirror asymmetry on average, and Braginskii's work [10] indicated that this was probably also necessary for large scale motions. Such asymmetry could probably most easily be provided, at least in part, by Coriolis forces, but again there are no quantitive results relating to any minimum speed of rotation.

While many dynamos suggested use differential rotation in the core as a major step in giving self-excitation, the differential rotation needed is probably only of the order of the observed westward drift, $10^{-6}$ that of the Earth's rotation, so again no useful lower limit is available.

Several « anti-dynamo » theorems have been proved. Unfortunately, despite a very common misconception, most of them refer to the geometry of the magnetic field, and not of the motions. The only specific restrictions on the motion are that it should not be purely toroidal, i.e. confined to spherical surfaces [5, 11] or purely radial [12]; this does not help us very much!

ROBERTS [6] has shown that a very wide variety of small scale motions would give dynamo action in a conducting fluid of infinite extent, and the work of CHILDRESS [13] suggests that this result will not be altered much if the conductor is bounded.

So we cannot at present set any lower limit to the speed of rotation of a planet if it is to produce a magnetic field, or specify at all closely the sorts of motion required in the conducting field.

## 6. – Estimation of the magnitude of core velocities.

Most work so far on all types of dynamo processes has been on the « kinematic dynamo problem »; a hopefully realistic motion is postulated, and any effect of the electromagnetic force on this motion is ignored. As in fact the electromagnetic (and Coriolis) forces are almost certainly much larger than the driving forces, this is a drastic oversimplification. Even so, the problem of the conditions in which such motions can be self-exciting is still far from solved.

Not all motions are self-exciting; for those which are there is a critical value of their speed at which there is stable excitation. Below this value any magnetic field would die away, while above it the induced field increases indefinitely (in practice the limit on available energy would limit the field produced, and the speed would remain at the critical value). From Maxwell's equations we obtain for the steady state

$$\nabla^2 \boldsymbol{H} = -4\pi\mu\sigma \nabla \times (\boldsymbol{v} \times \boldsymbol{H}) \,,$$

and when we make this equation nondimensional we find a nondimensional parameter $R_m$, the magnetic Reynold's number. $R_m = 4\pi\mu\sigma LV$ where $V$ is a typical speed of the motion and $L$ a typical length scale, and $\mu$ and $\sigma$ are the magnetic permeability and electrical conductivity of the fluid. (The factor $4\pi$ is appropriate for an unrationalized system of units; it is omitted for a rationalized system.)

The magnetic Reynold's number indicates the relative magnitude of the rate of production of magnetic field (in effect the induced e.m.f.s) and the rate of diffusion of the magnetic field (Ohmic dissipation). It follows that $R_m$ will have to be at least about unity if a finite field is to be maintained.

For Bullard-Gellman type motions the motions and the field have the same length scale, and there is no ambiguity in the definition of $R_m$. The critical value of $R_m$ is determined by the geometry of the particular motion; « efficient » motions have small values, and « inefficient » motions large values. In this steady-state case, the four physical parameters $\mu$, $\sigma$, $L$, $V$, occur only as their product; there is no way of separating their effects.

For small-scale motions there are two length scales, that of the motion and that of the main field. I will define $R_m$ in terms of the length scale of the main field; the ratio of the length scales is then another aspect of the geometry which affects the critical value of $R_m$.

It would obviously be helpful if there were a useful theoretical lower limit to critical $R_m$. Unfortunately the only limit which can be interpreted in terms of these overall parameters is one given by ROBERTS [14], and it is not very helpful. ROBERTS showed that for a conductor of infinite extent in which motions of maximum speed $V_{\max}$ were confined to a spherical region of radius $R$, then $R_m = 4\pi\mu\sigma R V_{\max}$ must at least equal unity. It is not known how removing the surrounding conductor will affect this value, but in any case it is almost certainly a gross underestimate of actual critical values of $R_m$.

We can therefore be guided only by the available empirical results. For a Bullard-Gellman type dynamo Lilley's [15] $V = 20$ corresponds to a value

of about 80 for a critical $R_m$ defined in terms of $V_{\max}$ and core radius. Ro-
BERTS [16] has also obtained a value of about 80 for a spherical dynamo with
purely axisymmetric motions. (P. H. ROBERTS (private communication) has
pointed out that both these values would be somewhat less if the velocities were
measured with respect to a co-ordinate system rotating at some mean velocity
of the core, rather than one based on zero rotation at the centre.) For their
laboratory models LOWES and WILKINSON [17] obtained a value of 200 based
on the radius of their rotating cylinders.

At present we do not know if the actual geomagnetic dynamo is likely to
be more or less « efficient » than these very simple models. However for large
motions a critical value $R_m = 100$ seems unlikely to be wrong by more than
a factor of about 3; this does give some constraint on core properties.

However, while on our present views the existence of a geomagnetic field
shows that a fluid conducting core must exist, only if we know the electrical
conductivity can we deduce a velocity.

We know $L = 3 \cdot 10^8$ cm and $\mu = 1$ e.m.u. For an iron core $\sigma$ has usually
been taken as $3 \cdot 10^{-6}$ e.m.u. [18] with a factor of 3 uncertainty. STACEY [19] suggested that the addition of moderate amounts of nickel and silicon could reduce this by a factor of 10, but the recent high-pressure measurements by
KEELER [20] on such alloys suggest a value of $(5 \div 10) \cdot 10^{-6}$ e.m.u. at the top
of the core. I will use $3 \cdot 10^{-6}$ as a typical, still very uncertain, value. Putting
$R_m = 100$ gives $V = 0.01$ cm(s)$^{-1}$. If we interpret the westward drift of the
geomagnetic field in terms of physical motion of the top of the core, then this
gives a speed of the order of 0.03 cm (s)$^{-1}$, so our estimate of $V$ is not unreasonable.

Another estimate of velocities in the core can be derived from the geomagnetic secular variation. Because of the short time scale of most of the
secular variation the lines of force leaving the core will effectively be tied to
the fluid particles at the top of the core. Using this « frozen field », infinite
conductivity, approximation attempts have been made [21-23] to extra-
polate $H$ and $\partial H/\partial t$ down to the core-mantle boundary, and hence to deduce
the velocity pattern at the surface of the core. Unfortunately the extrapolation is subject to considerable error; also BACKUS [24] has shown that it will
never be possible to determine $v$ unambiguously by this method. However
the magnitude of the deduced velocities is typically 0.01 cm/year, again consistent with the westward drift value.

## 7. – Hydromagnetic dynamo theory. Limitation on energy requirement.

To be more realistic a dynamo model should specify the driving forces,
and allow the electromagnetic (and Coriolis) forces to influence the motion.

However this problem is much more complicated and is very little understood; as yet there are no results relevant to the present discussion.

The discussion in the previous Section on possible limitation of $R_m$, the magnetic Reynold's number, is still applicable.

Obviously the magnitude of the magnetic field produced by a dynamo will depend on the power available, but without a detailed knowledge of the dynamo mechanism this dependence does not give any further information about the magnitude of the motions. Nor do we know if there is any minimum power requirement for self-excitation.

It is possible however to estimate the minimum power required to maintain the present field.

At the surface of the core the dipole field has magnitude 4 G at the pole, and the nondipole field contributes about as much again; these fields are probably of this magnitude throughout the core. In addition to these observable fields it is very likely that inside the core there are « toroidal » fields of order of magnitude 100 G; these fields have their lines of force confined to closed surfaces inside the conducting core, and come from the « winding up » of the lines of force of the dipole field by differential rotation in the core. Although they will « leak » slightly into the (slightly conducting) lower mantle we cannot observe them directly at the surface, and the accompanying Earth « currents » are almost certainly too small to detect [25]. Also it has been shown [26-28]; that the fields in the core will not have any significant effect on seismic waves. So it seems that at present there is no way of estimating the magnitude of the toroidal fields observationally.

To maintain the electric current system in the core giving the present dipole field requires about $3 \cdot 10^7$ W; to maintain a toroidal field of 100 G in the core requires about $10^{10}$ W. Even allowing for thermodynamic and mechanical inefficiencies these values are small compared with the total power dissipated in the Earth; there is however still dispute as to the actual mechanism which feeds power into the geomagnetic dynamo.

The energy source is probably either thermal convection (from radioactive heating), or the precession of the Earth's axis of rotation. So far most workers have assumed convection, but MALKUS [29] argues in favour of precession.

## 8. – Field reversals.

In any steady self-exciting dynamo the magnetic field can have either of two opposite directions, so the existence of both polarities in the history of the Earth's magnetic field is not surprising. Unfortunately practically nothing is known about possible instabilities of a geomagnetic dynamo.

In a series of papers RIKITAKE [30] and RIKITAKE and HAGIWARA [31] have attempted to investigate the stability and time variation of a Bullard-Gellman dynamo. However, because of the simplifications which had to be made their results must be suspect, particularly as GIBSON and ROBERTS [32] have shown that that particular Bullard-Gellman dynamo is almost certainly not self-exciting.

Because of the great difficulty in investigating the time variation of a reasonably realistic dynamo, BULLARD [33] suggested the investigation of a very simple lumped-constant, disc dynamo, and various authors have since done this on a two-disc model [34-39]. Although it is clear that such two-disc dynamos are unstable and oscillate and reverse, it is very doubtful if any quantitative results applicable to the Earth have been obtained.

The behaviour of such models depends very much on the value of a parameter which is the square root of the ratio of the mechanical acceleration time (in the absence of magnetic field) to the electromagnetic decay time (in the absence of motion). Both ALLAN [37] and LOWES [35] estimate this as being between $10^{-2}$ and $10^{-3}$ for a similar situation in the Earth. All the earlier published results were for values of this parameter of about unity. Because of computational difficulties no solutions have been obtained for values less than $10^{-2}$, but all the indications are that the solution would consist of periodic reversals of field with a period corresponding to a few years in the case of the Earth. The disc-dynamo model obviously involves too many unrealistic assumptions for any of its properties to be translated to the geomagnetic dynamo at present.

LOWES and WILKINSON [17] have investigated the behaviour of a laboratory self-exciting homogeneous dynamo. This has the mechanical simplicity of the Rikitake two-disc model, but with a three-dimensional current flow. It is unstable, and has various modes of oscillation, including periodic reversals. Unfortunately, in order to obtain a sufficiently high $R_m$ the model had to be made from a ferromagnetic, and they believe that the wave form and period of the reversals they find are determined primarily by the non-linearities of the ferromagnetic. Again it is unlikely that their results are relevant to the Earth's dynamo.

Thus at present there appears to be no way in which the increasing knowledge of the spectrum of geomagnetic reversals can be applied to increase our knowledge of the Earth's interior.

## 9. – Discussion.

It appears that with our present knowledge and understanding of the geomagnetic field we cannot add very much to the knowledge of the Earth's core derived from other aspects of geophysics. However, it is promising that the estimates so far possible are reasonable, and there is hope that they can be made more precise in the future.

## REFERENCES

[1] Quoted in E. C. Bullard and H. Gellman: *Phil. Trans. Roy. Soc.*, A **247**, 213 (1954).
[2] G. Backus: *Ann. of Phys.*, **4**, 372 (1958).
[3] A. Herzenberg: *Phil. Trans. Roy. Soc.*, A **250**, 543 (1958).
[4] F. J. Lowes and I. Wilkinson: *Nature*, **198**, 1158 (1963).
[5] See, for example, E. C. Bullard and H. Gellman: *Phil. Trans. Roy. Soc.*, A **247**, 213 (1954).
[6] G. O. Roberts: *Phil. Trans. Roy. Soc.*, A **266**, 535 (1970).
[7] See, for example, H. K. Moffatt: *Journ. Fluid Mech.*, **41**, 435 (1970).
[8] J. Larmor: *Report of the British Association Meeting 1919* (1920), p. 159.
[9] R. Hide: in *Magnetism and the Cosmos* (Edinburgh, 1967), p. 378.
[10] S. I. Braginskii: *Žurn. Éksp. Teor. Fiz*, **47**, 1084 (1964), English translation: *Sov. Phys. JETP*, **20**, 726 (1965).
[11] T. G. Cowling: *Quart. Journ. Mech. Appl. Math.*, **10**, 129 (1957).
[12] T. Namikawa and S. Matsushita: *Geophys. Journ.*, **19**, 395 (1970).
[13] S. Childress: in *The Application of Modern Physics to the Earth and Planetary Interiors* (London, 1969), p. 629.
[14] P. H. Roberts: *An Introduction to Magnetohydrodynamics* (London, 1967), p. 74.
[15] F. E. M. Lilley: *Proc. Roy. Soc.*, A **316**, 153 (1970).
[16] G. O. Roberts: *Numerical results on the dynamo action of axisymmetric motions in a sphere*, to appear.
[17] F. J. Lowes and I. Wilkinson: *Nature*, **219**, 117 (1968).
[18] E. C. Bullard: *Proc. Roy. Soc.*, A **197**, 433 (1949).
[19] F. D. Stacey: *Earth Planet. Sci. Lett.*, **3**, 204 (1967).
[20] R. N. Keeler: this volume, p. 188.
[21] A. B. Kahle, R. H. Ball and E. H. Vestine: *Journ. Geophys. Res.*, **72**, 4917 (1967).
[22] A. B. Kahle, E. H. Vestine and R. H. Ball: *Journ. Geophys. Res.*, **72**, 1095 (1967).
[23] R. H. Ball, A. B. Kahle and E. H. Vestine: *Journ. Geophys. Res.*, **74**, 3659 (1969).
[24] G. E. Backus: *Phil. Trans. Roy. Soc.*, A **263**, 239 (1968).
[25] P. H. Roberts and F. J. Lowes: *Journ. Geophys. Res.*, **66**, 1243 (1961).
[26] L. Knopoff: *Journ. Geophys. Res.*, **60**, 441 (1955).
[27] E. A. Kraut: *Journ. Geophys. Res.*, **70**, 3927 (1965).
[28] F. E. M. Lilley and D. E. Smylie: *Journ. Geophys. Res.*, **73**, 6527 (1968).
[29] W. V. R. Malkus: see, for example, *The Geodynamo*, this volume, p. 38.
[30] T. Rikitake: *Bull. Earthquake Res. Inst. Tokyo*, **33**, 571 (1955); **34**, 283 (1956); **37**, 245 (1959).

[31]  T. RIKITAKE and Y. HAGIWARA: *Journ. Geomag. Geoelec.*, **18**, 393 (1966); **20**, 57 (1968).

[32]  R. D. GIBSON and P. H. ROBERTS: in *Magnetism and the Cosmos* (Edinburgh, 1967), p. 108.

[33]  E. C. BULLARD: *Proc. Cambridge Phil. Soc.*, **51**, 744 (1955).

[34]  T. RIKITAKE: *Proc. Cambridge Phil. Soc.*, **54**, 89 (1958).

[35]  F. J. LOWES: unpublished.

[36]  D. W. ALLAN: *Nature*, **182**, 469 (1958).

[37]  D. W. ALLAN: *Proc. Cambridge Phil. Soc.*, **58**, 671 (1962).

[38]  J. H. MATHEWS and W. K. GARDNER: U.S. Naval Res. Lab. Rept. 5886 (1963).

[39]  A. E. COOK and P. H. ROBERTS: *Proc. Cambridge Phil. Soc.*, **63**, 547 (1970).

# Motions in the Fluid Core.

W. V. R. Malkus

*Department of Mathematics, M.I.T. - Cambridge, Mass.*

## 1. – The geodynamo.

As outlined by Lowes in the preceding lecture, the geomagnetic field is presumed to be due to motions in the Earth's fluid core. Doubts concerning the possibility of dynamo action in a homogeneous electrically conducting fluid have been removed by the theoretical construction of velocity fields which exhibit magnetic instability. However, the origin and character of the motions responsible for the Earth's dynamo have not been determined. That is to say, the elementary kinematic problem has been resolved while the numerous dynamic problems have barely been touched. It is generally believed that heat sources in the core produce convective motions which in turn drive the geodynamo. Convection is a theoretically attractive mechanism because of its possible applicability to stellar dynamos as well as to the Earth. However, it has been difficult to find a heat source sufficient to meet even the minimum energy needs of a geodynamo and compatible with surface observations of heat flux. Hence, in the second of these four lectures, we will discuss flow due to the precession of the earth. It is my belief that precession is the most likely cause of the motions which drive the geodynamo.

We will first discuss the foundations of the kinematic theory.

The first and most obvious suggestion of the geomagnetic data is that we are looking at a process; not a static thing, not a remnant of the past. One infers the existence of a mechanism involved in transferring energy from fields of motion to magnetic fields to Ohmic dissipation.

Let us first recall what a dynamo is. The simplest example is the « homopolar dynamo », shown in Fig. 1. There, a simple disc of metal is rotated about its axis in the presence of a magnetic field, $B$ imposed along that axis. As a consequence, a potentual gradient $v \times B$ occurs in the plane of the disc, where $v$ is the local velocity of the disc. If the charge accumulated at the periphery of the disc is permitted to flow through a coil back to the axis, then the resulting field can re-inforce the originally imposed field **B**. This system becomes a dynamo when this induced field becomes equal to the field required to produce it. The criterion of this critical behaviour is determined by the non-dimensional combination

$$(1.1) \qquad\qquad R_m \equiv \mu\sigma|\boldsymbol{v}|L \gg 1\,,$$

where the magnetic Reynolds number, $R_m$, is defined as the product of the magnetic permeability $\mu$, the conductivity $\sigma$, some measure of the velocity $|\boldsymbol{v}|$,

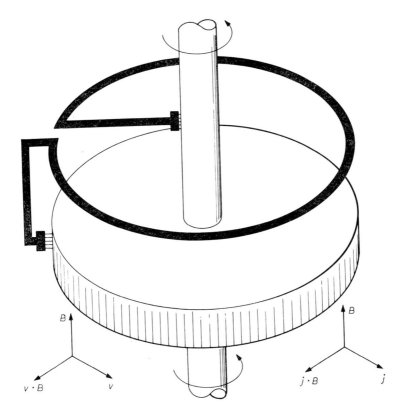

Fig. 1. – The homopolar disc dynamo.

some appropriate scale length $L$, of the dynamo system. When this number is large compared to 1, regeneration can occur. Note, however, that the homopolar machine is not a simply-connected dynamo. It requires complicated conduction paths, and splipping brushes in order to operate. It was not clear that a homogeneous conducting fluid could be put into motion in such a way as to initiate this dynamo action.

We first ask the question «Under what conditions can a homogeneous conducting fluid maintain a magnetic field?» The Maxwell equations, with the displacement current neglected, are written:

(1.2) $$\nabla \times \boldsymbol{H} = \boldsymbol{j} ,$$

(1.3) $$\nabla \times \boldsymbol{E} = -\frac{\partial \boldsymbol{B}}{\partial t} ,$$

(1.4) $$\nabla \cdot \boldsymbol{B} = 0 ,$$

(1.5) $$\nabla \cdot \boldsymbol{D} = 0 ,$$

where $\boldsymbol{B} = \mu \boldsymbol{H}$, $\boldsymbol{H}$ is the magnetic field strength, $\boldsymbol{D} = \varepsilon \boldsymbol{E}$, $\boldsymbol{E}$ is the electric field strength and $\varepsilon$ is the dielectric constant. The current density is:

(1.6) $$\boldsymbol{j} = \sigma(\boldsymbol{E} + \boldsymbol{v} \times \boldsymbol{B}) \qquad \left( \text{for } \frac{v^2}{c^2} \ll 1 \right).$$

Taking the curl of the eq. (1.2) and using eqs. (1.3) and (1.6), one can write the magnetic diffusion equation

(1.7) $$\frac{\partial \boldsymbol{H}}{\partial t} - \frac{1}{\mu\sigma} \nabla^2 \boldsymbol{H} = \nabla \times \boldsymbol{v} \times \boldsymbol{H} .$$

If $\boldsymbol{v}$ is assumed to be a known function, eq. (1.7) is a homogeneous equation of second order with nonconstant coefficients. Presumably, $\boldsymbol{H}$ can have an exponential instability for certain classes of these nonconstant coefficients. Among the first explorers of this problem was COWLING, who established that no axi-symmetric $\boldsymbol{H}$ field could be maintained by dynamo action, no field for which the radial component of velocity was zero could give rise to dynamo action and no two-dimensional velocity fields could give rise to dynamo action. Hence, there was a period of time when it was very doubtful that homogeneous dynamos were possible.

This difficulty was resolved by BACKUS and HERZENBERG. Herzenberg's model was particularly simple and is exhibited in Fig. 2. Two spheres of conducting material are rotated in a surrounding medium of the same conductivity. One can see that the symmetric currents

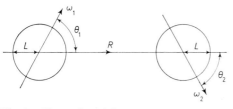

Fig. 2. – Herzenberg's homogeneous dynamo.

which would flow due to a magnetic field oriented along $\omega_1$, will not sustain a dynamo action in the rotor $\omega_1$. However, the axisymmetric magnetic field that would result can act to excite currents in the distant rotor $\omega_2$. It is this coupling between the two separated induction fields that permits dynamo action The criterion for dynamo action to first order in the separation distance $R$ is:

(1.8)         $[(\frac{1}{5} R'_m \sin \theta_1 \sin \theta_2 \sin \varphi)^2 (\cos \theta_1 \cos \theta_2 - \sin \theta_2 \cos \varphi) - 3]^2 = 0$

where $R'_m = R_m(L/R)^3$, $\theta_{1,2}$ are indicated on Fig. 2, $L$ is the radius of each rotor and $\theta$ is the angle between the $\boldsymbol{\omega}_1\text{-}\boldsymbol{R}$ plane and the $\boldsymbol{\omega}_2\text{-}\boldsymbol{R}$ plane.

A laboratory model of this process constructed by LOWES and WILKINSON not only exhibited intense dynamo action above a critical angular velocity, but also had field reversals of long period. The field reversals were related to the strong back-coupling to the magnetic and velocity fields arising from the Lorentz force $\boldsymbol{j} \times \boldsymbol{H}$. As the magnetic instability grew it slowed down the rotating spheres below the critical angular velocity. This caused the field to decay, the spheres to accelerate, and the process to repeat itself. A satisfactory theory for this dynamic aspect of a Herzenberg dynamo has yet to emerge.

With the question regarding kinematic homogeneous dynamo action behind us, the next problem is the determination of a « natural » flow process which can act as a dynamo. For example, can convection due to radial temperature contrasts in a rotating sphere cause a dynamo?—and what does the rotation have to do with it?

It is generally believed that the answer to the first part of this question is « yes », but it is not yet proved. An answer to the second part of the question may very well be the key to understanding the principal dynamic role of magnetic fields in nature. Many investigators have concluded that the magnitude of the Lorentz force is comparable to the Coriolis force in the Earth and in the sun. Almost as many different reasons for this gross balance have been advanced. Since magnetic fields do not directly transfer heat, it is my view that the Lorentz force arises in conducting fluids to relax the gyroscopic constraints on motion due to the rotation, hence assisting in the release of available potential energy. This view is supported in a 1959 paper on initial finite-amplitude magnetoconvection. However, a definitive study has yet to be made.

The actual energy sources for motion in the Earth's core remain uncertain. If they are thermal, then the total dissipation rate integral requires that the heat flux from the core to the mantle be more than a factor of ten greater than the Ohmic dissipation rate. VERHOOGEN (1961) proposed that solidification of the inner core could supply an adequate heat flux to maintain core convection under the condition that the density contrast between liquid and solid was very small. Pending better determinations of the physical properties of the core fluid, the latent heat of solidification appears to be the most plausible energy source for a convection-driven geodynamo.

## BIBLIOGRAPHY OF SECTION 1

W. M. ELSASSER: *Phys. Rev.*, **55**, 489 (1939).
W. M. ELSASSER: *Rev. Mod. Phys.*, **22**, 1 (1950).
A. HERZENBERG: *Phil. Trans. Roy. Soc. London*, A **250**, 543 (1958).
R. HIDE and P. H. ROBERTS: *Physics and Chemistry of the Earth*, Chap. 4 (London, 1961).
F. J. LOWES and I. WILKINSON: *Nature*, **198**, 1158 (1963).
W. V. R. MALKUS: *Astrophys. Journ.*, **130**, 259 (1959).
J. VERHOOGEN: *Geophys. Journ.*, **4**, 276 (1961).
G. VERONIS: *Journ. Fluid Mech.*, **5**, 401 (1959).
G. E. BACKUS: *Ann. der Phys.*, **4**, 372 (1958).
E. C. BULLARD: *Monthly Notices Roy. Astron. Soc. Geophys. Suppl.*, **5**, 248 (1948).
E. C. BULLARD: *In the Earth as a Planet* (Chicago, 1954).
T. G. COWLING: *Montly Notices Roy. Astron. Soc.*, **94**, 39 (1933).

## 2. – Precessional torques and geomagnetism.

In our first Section, aspects of the kinematic geodynamo problem were explored. In a brief discussion of convective dynamos, the difficulty of finding a sufficient thermal energy source in the Earth's interior was noted.

In this Section, I wish to describe for you the classical idealization of flow due to precession, certain experiments in which some aspects of this flow were realized, and the hydromagnetic implications of these observations. In a fourth Section, the detailed study of the precessional problem will be taken up again and we will find that certain of the newly observed fluid-dynamical features of precessional motion can be interpreted, utilizing the mathematical tools of singular perturbation theory.

Of the several mechanisms for inducing motion in the core that have been proposed, the precession of the Earth is the only one whose magnitude and character are well known. The Earth precesses with a period of 25,800 years, due to the gravity fields of the moon and the sun acting on the Earth's equatorial bulge. Figure 3 illustrates the aspects of Earth structure and average motion that are relevant to this study. The precession vector $\Omega = 7.71 \cdot 10^{-12}$ radian per second and is normal to the plane of the ecliptic. The Earth's angular rotation vector $\omega = 7.29 \cdot 10^{-5}$ radian per second and is inclined at 23.5 degrees relative to $\Omega$. The lightly shaded zone in Fig. 3 is the Earth's molten core, of mean radius $R = 3.47 \cdot 10^8$ cm. The mean radius of the Earth as a whole, $R_1$, is roughly $2R$, while the radius, $R_2$, of the presumably solid inner core is roughly $0.4R$. The rotation of the Earth causes an equatorial bulge, resulting in a difference between the moments of inertia about the polar and the equatorial axes. The ratio of this difference to the moment of inertia about the polar axis is known as the dynamic ellipticity. The dynamic ellipticity of the

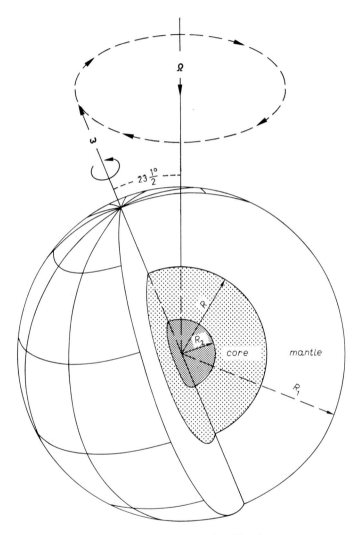

Fig. 3. – The precessing Earth.

Earth as a whole is $3.28 \cdot 10^{-3}$, and that of the core alone is $2.45 \cdot 10^{-3} \pm 2$ percent. The dynamic ellipticity of the core is three-quarters that of the mantle, due to the core's greater density (approximately 10 g/cm³).

The precession rate of a planetary body is directly proportional to its dynamic ellipticity, but independent of its radius or mass. Hence, if the core and mantle were not coupled, the core would precess at only three-quarters the rate of the mantle. Thirty thousand years of such uncoupled precession would lead to relative velocities of $10^4$ cm/s at the core-mantle boundary. Of course, core and mantle are coupled by torques resulting from their relative motion. These torques will increase until core and mantle precess at the same average rate.

The idealized flow indicated by the relative precession of core and mantle was first investigated by POINCARÉ and HOUGH in the last century. They described the flow of the fluid inside a precessing spheroidal container of major radius $R$.

The equations appropriate to these descriptions, first for the continuity of a mass

$$(2.1) \qquad\qquad \nabla \cdot \boldsymbol{v} = 0 ,$$

second for the continuity of momentum

$$(2.2) \qquad \frac{\partial \boldsymbol{v}}{\partial t} - E\nabla^2\boldsymbol{v} + 2\boldsymbol{\Omega} \times \boldsymbol{v} + \boldsymbol{v} \cdot \nabla\boldsymbol{v} + \nabla P = 0 ,$$

are here scaled so that the velocity on the surface $S$ of the container is written as

$$(2.3) \qquad\qquad \boldsymbol{v} = \boldsymbol{k} \times \boldsymbol{r} ,$$

where $\boldsymbol{k}$ is the unit vector along the axis of rotation and $\boldsymbol{r}$ is the position vector measured from the center of the system. The scaling is also such, that the viscous term appears multiplied by the Ekman number,

$$(2.4) \qquad\qquad E = \frac{\nu}{\omega R^2} ,$$

where $\nu$ is the kinematic viscosity, $\omega$ the angular velocity of rotation, and $R$ the radius of the container. The shape of the container is given by

$$(2.5) \qquad\qquad r^2 + \eta(\boldsymbol{k} \cdot \boldsymbol{r})^2 = 1 ,$$

where $\eta$ is a measure of the ellipticity of the oblate spheroid.

The classical problem treated only one aspect of the general boundary condition, eq. (2.3), both POINCARÉ and HOUGH presumed that the boundary condition

$$(2.6) \qquad\qquad \boldsymbol{v} \cdot \boldsymbol{n} = 0 \text{ on } S ,$$

where $n$ is a unit vector normal to surface $S$, would be sufficient to describe all but subtle viscous aspects of the problem. A generalized solution to the problem with this simple boundary condition is easily shown to be one of constant vorticity, given by

$$(2.7) \qquad \boldsymbol{v} = \boldsymbol{\omega}_f \times \boldsymbol{r} + \nabla A \;,$$

when this velocity field is substituted into the original equations, the solution for the vorticity of the fluid is given by:

$$(2.8) \qquad \boldsymbol{\omega}_f = \alpha \boldsymbol{k} - \frac{\alpha(2+\eta)(\boldsymbol{\Omega} \times \boldsymbol{k})x\boldsymbol{k}]}{\alpha\eta + 2(\boldsymbol{k}\cdot\boldsymbol{\Omega})(1+\eta)} \;,$$

and for the potential term needed to satisfy the oblate spheroidal boundary conditions by

$$(2.9) \qquad A = \frac{-\alpha\eta}{\alpha\eta + 2(\boldsymbol{k}\cdot\boldsymbol{\Omega})(1+\eta)} \, (\boldsymbol{\Omega} \times \boldsymbol{k})\cdot\boldsymbol{r}(\boldsymbol{k}\cdot\boldsymbol{r}) \;.$$

As one might have anticipated this theory of the flow is not unique and contains an arbitrary parameter $\alpha$. When $\alpha = 1$, as was assumed by POINCARÉ and HOUGH, then $\boldsymbol{\omega}_f$ approaches the angular velocity of the container as $\boldsymbol{\Omega}$ approaches zero. What is fascinating about this solution is that it is an exact nonlinear solution of the entire problem, except for a small tangential component of the velocity at the boundary. If one defines a relative amplitude, $\varepsilon$, as

$$(2.10) \qquad \varepsilon^2 = (\boldsymbol{\omega}_f - \boldsymbol{k})^2 \;,$$

then in the linear theory, the order of magnitude of $\varepsilon$ is given by:

$$(2.11) \qquad \varepsilon = O\left(\frac{|\boldsymbol{\Omega} \times \boldsymbol{k}|}{\eta}\right) .$$

The resulting flow is one of smooth elliptical stream lines around an axis, not quite coincident with the axis of rotation of the spheroidal shell. It was believed by early investigators that this flow was too smooth and too uniform, to provide any dynamo action.

A recent investigation of these Poincaré-like flows by STEWARTSON and ROBERTS included a viscous correction in the linear theory for the small inadequacy of this theory on the boundary surface. They concluded that the viscous correction would lead to a thin boundary layer, but that the principal flow would remain unaltered.

I constructed an experiment, using a precessing oblate spheroidal cavity, to test these conclusions in the laboratory. This experiment exhibited dramatic departures from the laminar Poincaré flow. Figure 4 illustrates this study. A plastic spheroid, of major axis 25 cm and minor axis 24 cm, was filled with water and rotated at 60 revolutions per minute around its minor axis. The

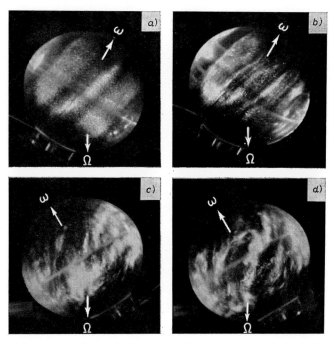

Fig. 4. – Precession in a laboratory spheroid: a) laminar zonal flow, b) initial instability, c) and d) rotationally constrained turbulence.

spheroid, its supports, and its motor were mounted on a horizontal, rotatable table, with the minor axis of the spheroid inclined at 30 degrees to the vertical. A pinch of aluminium powder dispersed in the water was illuminated by a beam of light normal to the rotation axis, permitting visualization of the process. Rotation of the table at $\frac{3}{4}$ revolution per minute caused the steady two-dimensional flow seen in Fig. 4 a). This steady flow already exhibits a departure from the general Poincaré solution, for the light and dark bands produced by the aluminium particles suggests shear zones of considerable magnitude in the toroidal flow in the container. Rotation of the table at 1 revolution per minute led to the wavelike instabilities of these toroidal flow seen in Fig. 4 b), while rotation at $\frac{4}{3}$ revolution per minute caused rather dramatic turbulent flows seen in Fig. 4 c) and d).

A program of measurement was undertaken to determine the flow field prior to instability, the critical parameters determining instability, and the motor torques needed to sustain the flow. In a fourth Section, the theoretical description of these various regimes of flow will be assessed. At this point I wish to explore the implications of these disordered flows as an energy source for a dynamo process.

Perhaps the most straightforward approach to the problem is to note that the transformation of co-ordinates from inertial space to a uniformly rotating and precessing space leads to the usual centrifugal and Coriolis forces plus an additional force. This unfamiliar precessional force per unit volume is written

$$(2.12) \qquad [(\boldsymbol{\omega} \times \boldsymbol{\Omega}) \times \boldsymbol{r}] \,,$$

I will refer to this force as the Poincaré force, in honor of his early studies of precession. The Poincaré force is akin to the usual centrifugal force, but since it produces a net torque it cannot be balanced by pressure alone. Due to the difference in core and mantle ellipticity, one-fourth of the Poincaré force remains unbalanced in the core and produces fluid motion.

A fluid velocity $V$ in the core will give rise to the usual Coriolis force per unit mass, $2\boldsymbol{\omega} \times \boldsymbol{V}$. Such electric currents as may be associated with this flow will produce the Lorentz force per unit mass $(\mu/\varrho)(\boldsymbol{j} \times \boldsymbol{H})$, where $\varrho$ is the density of the fluid, $\mu$ is the magnetic permeability, $\boldsymbol{H}$ is the magnetic field, and $\boldsymbol{j}$ is the electric current density. Here, and in what follows, we used unscaled variables and Gaussian electromagnetic units throughout.

I assume, tentatively, that the Poincaré force in the earth's core is large enough to cause the turbulent type of flow seen in the experimental situation, but presumably magnetic turbulence. I will define a fully magnetic turbulent flow as meaning that a gross balance has been struck between the fluctuating Coriolis and Lorentz forces in the interior of the flow and that each of these fluctuating responses is of the same magnitude as the steady Poincaré driving force. Forces due to pressure gradients can be as important as the Coriolis and Lorentz forces, but I will show that it is sufficient to consider only the two major inertial forces and the electromagnetic force in order to obtain an estimate of the velocities and magnetic fields which would be associated with such flow.

This gross balance is written:

$$(2.13) \qquad \frac{1}{4} |(\boldsymbol{\omega} \times \boldsymbol{\Omega}) \times \boldsymbol{r}| \simeq |2\boldsymbol{\omega} \times \boldsymbol{V}| \simeq \left| \frac{1}{4\pi\varrho} (\nabla \times \boldsymbol{H}) \times \boldsymbol{H} \right| \,.$$

Hence the estimate for the magnitude of a characteristic velocity $V$ is:

$$(2.14) \qquad V \simeq \frac{|\boldsymbol{\omega} \times \boldsymbol{\Omega}|}{8|\boldsymbol{\omega}|} R \simeq \frac{4}{3} \cdot 10^{-4} \text{ cm/s} \,.$$

For scales of motion comparable to the radius of the core, $\nabla$ in eq. (2.13) is replaced by $1/R$, and the estimate for the magnitude, $H$, of a characteristic magnetic field is:

$$(2.15) \qquad\qquad H \simeq (\pi\varrho\,|\boldsymbol{\omega}\times\boldsymbol{\Omega}|R^2)^{\frac{1}{2}} \simeq 29 \text{ gauss} .$$

The vector character of the force balance (eq. (2.13)), requires that the estimated magnetic field $H^2$ be interpreted as the product $H_{\mathrm{p}}H_{\mathrm{T}}$, where $H_{\mathrm{p}}$ is the magnitude of the poloidal components of the field and $H_{\mathrm{T}}$ is the magnitude of the toroidal component of the field. The poloidal components lie in meridian planes containing the axis of rotation. The toroidal component is at right angles to these planes. Similarly, the characteristic velocity $V$ should be interpreted as representative of the geometric mean of poloidal and toroidal components. A balance of the average radial Coriolis force and the average radial Lorentz force is not included in the estimates given above, but is used in the following section. Neither $V$ nor $H$ are directly observable quantities, since they are estimates for fields within the molten core material.

The magnitude of the Earth's dipole field at the core-mantle boundary, as inferred from surface observations, is approximately 5 gauss. In order to determine the external dipole associated with the internal field $H$, one must establish how much of $H$ can « leak out » of the electrically conducting mantle. A gross balance of the terms in the magnetic-diffusion equation provides such an estimate, but this estimate depends explicitly upon the value of the core conductivity.

This balance between the loss of magnetic field by ohmic dissipation and its regeneration by the fluid motion is

$$(2.16) \qquad\qquad \left| \frac{1}{4\pi\sigma} \nabla\times(\nabla\times\boldsymbol{H}) \right| \simeq |\nabla\times(\boldsymbol{V}\times\boldsymbol{H})| ,$$

where $\sigma$ is the electrical conductivity of the fluid. For scales of motion comparable to the radius of the core, the balance of eq. (2.16) may be interpreted as the estimate

$$(2.17) \qquad\qquad H_0 \simeq \frac{1}{4\pi\sigma R}\frac{H}{V} = \frac{2|\boldsymbol{\omega}|}{\sigma R}\left(\frac{\varrho}{\pi|\boldsymbol{\omega}\times\boldsymbol{\Omega}|}\right),$$

where $H_0$ is defined as the magnitude of the magnetic field at right angles to the velocity $V$ and is presumed to be small compared to $H$, and where the ratio $H:V$ is espressed in terms of the precession rate through the use of eqs. (2.14) and (2.15). Since the average fluid velocity must be parallel to the core-mantle boundary, $H_0$ is also an estimate for the magnetic field external to the core. A mean value for the electrical conductivity of the core material, determined from the extrapolation of laboratory measurements, is $7 \cdot 10^{-6}$ (abohm centimeter)$^{-1}$. However, an uncertainty of at least a factor of 3, in either direction, is given in the literature. With this choice for $\sigma$, the estimate (eq. (2.17)) for the external magnetic field at the core-mantle boundary is

$$H_0 \simeq 7 \text{ gauss} \pm 300 \% .$$

If the estimate for $H_0$ had led to a value equal to or larger than $H$, the appropriate interpretation of the dissipation-regeneration balance equation would have been that the magnetic field could not be maintained by the velocity field and would vanish. This would occur if the precession $|\Omega|$ where 1/16 its present value, or if the conductivity $\sigma$ where 1/4 the mean value used in arriving at the estimate given above.

Having found an approximate value for $H_0$, we can make a secondary estimate of the internal fields. Since the estimate value for $H_0$ is also an estimate for one component of the poloidal field just inside the core-mantle boundary, then $H^2 \equiv H_p H_T \simeq H_0 H_T$. From eqs. (2.15) and (2.17), we obtain the relation

$$(2.18) \qquad H_T \simeq \frac{H^2}{H_0} = \frac{\sigma}{2|\omega|} \varrho^{\frac{1}{2}} \pi |\boldsymbol{\omega} \times \boldsymbol{\Omega}|^{\frac{1}{2}} R^3 \simeq 120 \text{ Gauss} \pm 300 \% .$$

Finally, a secondary estimate for a mean value of the toroidal velocity field, $V_T$, follows from the balance, in eq. (2.13), of the radial Coriolis force and the radial Lorentz force. This relation is written:

$$(2.19) \qquad V_T = \frac{H_T^2}{8\pi\varrho|\omega|R} \simeq \frac{(\sigma\pi)^2 R^5}{32} \left( \frac{|\boldsymbol{\omega} \times \boldsymbol{\Omega}|}{|\boldsymbol{\omega}|} \right)^3 \simeq 2.3 \cdot 10^{-3} \text{ cm/s} \pm 1000 \% ,$$

where eq. (2.18) is used to express $H^2$ in terms of the precession rate and the conductivity. Order-of-magnitude accuracy is the most one might expect from a secondary estimate such as this. If $V_T$ is interpreted as representative of the westward-drift component of core surface velocity, then the value in eq. (2.19) is one-sixth the mean drift velocity inferred from recent geomagnetic data.

The energy consumption of the precession-driven geodynamo can be found directly from the earlier estimates. The characteristic stress acting on the core-mantle boundary must equal the momentum transport just inside the boundary. From eq. (2.15) we find this momentum transport to be

$$(2.20) \qquad \frac{H_p H_T}{4\pi} = \frac{\varrho}{4} \left( |\boldsymbol{\omega} \times \boldsymbol{\Omega}| R^2 \right) = 68 \text{ dyne/cm}^2 .$$

An estimate of the total work, $W$, done by the mantle on the core is then, from eqs. (2.19) and (2.20),

$$(2.21) \qquad W = \left( \frac{H_p H_T}{4\pi} \right) V_T (4\pi R^2) \simeq \frac{\varrho \sigma^2 \pi^3 R^9 |\boldsymbol{\omega} \times \boldsymbol{\Omega}|^4}{32 |\boldsymbol{\omega}|^3} = 2.3 \cdot 10^{17} \text{ erg/s} \pm 1000\%,$$

and this would be consumed in ohmic heating primarily. For comparison, note that the estimated rate of dissipation due to tidal interaction of earth, moon, and sun is $3 \cdot 10^{19}$ ergs per second. Like the tidal dissipation process, the energy for the precession-driven geodynamo must come from the kinetic energy stored in the earth's rotation. Unlike the tidal process, the response of the core fluid to precessional forces does not produce a reaction torque on the moon or sun. Hence, in order for the total angular momentum of the Earth, moon, sun system to be conserved, only the rotational energy in the nonconserved component of the Earth's rotation can supply the dynamo. This is the component of rotation in the plane of the ecliptic, the component at right angles to $\Omega$ in Fig. 3. It is sufficient at present to maintain the geomagnetic field for many Earth lifetimes. However, in an earlier eon when the moon was considerably closer to the Earth than it is now, the dynamo dissipation might well have exceeded the total tidal dissipation and would have contributed significantly to internal heating in the Earth.

It is rather fortunate, that a scale analysis should lead to unique qualitative results as does the scale analysis starting with eq. (2.13). It is then possible to test, probably within an order of magnitude, the suggestion that precessional torques can be responsible for the dynamo process. One might have concluded from such an analysis that the physical parameters of the process could not possibly have led to a dynamo. One might have concluded that the physical parameters of the problem would have led to a dynamo many, many orders of magnitude greater than the one we presume exists in the Earth's interior. That the analysis leads to values within the range of the real geodynamo strongly supports the suggestion that the geodynamo has this energy source.

## BIBLIOGRAPHY OF SECTION 2

S. S. Hough: *Phil. Trans. Roy. Soc. London*, **186**, 469 (1895).

H. S. Jones: in *The Earth as a Planet*, Chap. 1 (Chicago, 1954).

G. J. MacDonald: in *Advances in Earth Science*, Chap. 4 (Cambridge, 1966).

W. V. R. Malkus: *Journ. Geophys. Res.*, **68**, 2871 (1963).

W. V. R. Malkus: *Science*, **160**, 259 (1968).

H. Poincaré: *Bull. Astron.*, **27**, 321 (1910).

P. H. Roberts and K. Stewartson: *Astrophys. Journ.*, **137**, 777 (1963).

## 3. – Hydromagnetic planetary waves.

In a previous Section we discussed the secular variations of the Earth's magnetic field which are believed to be manifestations of the dynamo process which maintains that field. The various harmonic components of the secular variation appear to move primarily towards the west. HIDE (1967) has recently proposed that these hydromagnetic flows are analogous to the two-dimensional Rossby or $\beta$-plane waves, which are studied in meteorology and oceanography. I proposed in a similar study, that two-dimensional hydromagnetic waves can develop on toroidal shear layers in a rotating sphere. In both of these studies these two-dimensional waves moved slowly to the west.

However, the observed secular variations show negligible evidence of two-dimensionality. If the dynamo is as turbulent as these secular variations suggest, it seems likely that there are in the core many motions of a three-dimensional nature. If there are primarily westward-moving waves, this may give us some significant information about the mechanical processes which produce them.

Today, I wish to construct for you a suitable idealization in which the various modes of hydromagnetic oscillation of a rotating spheroid can be determined. This will be a combination of the free mechanical modes that can exist due to rotation of a fluid (a concept which may be unfamiliar to some of you) and Alfvén-like disturbances in a bounded region (which will probably appear somewhat more familiar). By good fortune the choice of a uniform electric current density to define the basic magnetic field leads to a class of eigenvalue problems which have been first studied by POINCARÉ (1910). It will take us some time to reach that point in the analysis, but, when we finally do we will be able to derive several dispersion relations from the general eigenvalue equation to establish that in the hydromagnetic case, the free model have phase velocities both to the east and west.

This basic magnetic filed has been selected with two criteria in mind. The first of these that it be a field reflecting the gross structure of the presumed field in the earth's core. The second criterion is that the basic state be stable, hence, realizable in an experiment.

Various authors have studied the stability to axisymmetric disturbances of toroidal magnetic fields in rotating incompressible fluids. This problem is akin to the classical study by TAYLOR on instabilities between rotating cylinders. The appropriate descriptive equations are:

(3.1) $$0 = (\mathrm{d}\boldsymbol{V}/\mathrm{d}t) - \nu\nabla^2\boldsymbol{V} + \nabla P/\varrho + \boldsymbol{H}\times(\nabla\times\boldsymbol{H}) + 2\boldsymbol{\omega}\times\boldsymbol{V}\,,$$

(3.2) $$0 = \nabla\cdot\boldsymbol{V} = \nabla\cdot\boldsymbol{H}\,,$$

(3.3) $$0 = (\partial\boldsymbol{H}/\partial t) - \eta\nabla^2\boldsymbol{H} - \nabla\times(\boldsymbol{V}\times\boldsymbol{H})\,,$$

where $P$ is the pressure, $\varrho$ the costant density, $V$ the vector velocity, $\nu$ the kinematic viscosity, $\boldsymbol{\omega}$ the angular velocity of rotation, $\eta$ the magnetic diffusivity and

(3.4) $$H = (\mu/4\pi\varrho)^{\frac{1}{2}} \mathcal{H} ,$$

where $\mathcal{H}$ is the magnetic filed and $\mu$ the constant permittivity of the medium.

It has been established (PAO, 1966) that the marginal condition for stability in the presence of diffusive processes is a state of uniform current density parallel to the axis of rotation.

We can anticipate that conditions close to the marginal state are achieved within the Earth, just as one anticipates that thermal turbulence in a planetary or stellar atmosphere produces a state of an adiabatic lapse rate. However, the Earth will also have a boundary region where the toroidal field almost vanishes, for the geodynamo will close most of its current paths within the highly conductive core. These nonlinear boundary regions will not be tackled today and therefore we will restrict the applicability of our wave study to scales which are large compared to the scale of the boundary region.

We may imagine the uniform current is imposed on the rotating fluid by an external potential. Solutions of the hydromagnetic equations are now sought for $\nu = \eta = 0$ and no initial velocity field.

In linearized from eqs. (3.1, (3.2) and (3.3) are written

(3.5) $$0 = (\partial V/\mathrm{d}t) + \nabla P' + H_0 \times (\nabla \times H) + H \times (\nabla \times H_0) + 2\boldsymbol{\omega} \times V ,$$

(3.6) $$\nabla \cdot V = 0 = \nabla \cdot H ,$$

(3.7) $$\partial H/\partial t = \nabla \times (V \times H_0) ,$$

(3.8) $$P' = P/\varrho + \tfrac{1}{2} H_0^2 ,$$

where $\nabla \cdot H_0 = 2j$ is a constant vector parallel to $\boldsymbol{\omega}$, and we choose

(3.9) $$H_0 = j \times r ,$$

where $r$ is the position vector measured from the axis of rotation.

The conditions on $V$ and $H$ at the spherical boundary are that

(3.10) $$V \cdot n = 0 = H \cdot n ,$$

where $n$ is a unit vector normal to the surface.

The boundary condition on $H$ is a consequence of eq. (3.7) and the $V$ boundary condition.

None of the coefficients of eqs. (3.5), (3.6) and (3.7) depend upon the longitude angle $\varphi$ or time. If boundary conditions are chosen which are also independent of $\varphi$ and time, we may seek solutions of the form

$$(3.11) \qquad R(r, \varphi, z, t) = Q(r, z) \exp\left[i(k\varphi - \sigma\omega t)\right],$$

where $R$ stands for any of the variables $V_r$, $V_\varphi$, $V_z$, $H_r$, $H_\varphi$, $H_z$ and $P'$. In eq. (3.11) the cylindrical co-ordinates $r$, $\varphi$, $z$ are used with $z$-axis parallel to $\omega$; $k$ indicates the zonal wave-number, and $\sigma$ is a nondimensional frequency in units of $\omega$.

As a consequence of eqs. (3.9) and (3.11)

$$(3.12) \qquad \nabla \cdot (\boldsymbol{V} \cdot \boldsymbol{H}_0) = [\hat{r}(\partial V_r/\partial\varphi) + \hat{\varphi}(\partial V_\varphi/\partial_\varphi) + \hat{z}(\partial V_z/\partial\varphi)] = ikj\boldsymbol{V},$$

where $\hat{r}$, $\hat{\varphi}$, $\hat{z}$ are unit vectors in the indicated directions and $j$ is the scalar magnitude of $\boldsymbol{j}$. Also it follows that

$$(3.13) \qquad \boldsymbol{H}_0 \times (\nabla \times \boldsymbol{H}) = -ikj\boldsymbol{H} + \nabla(\boldsymbol{H} \cdot \boldsymbol{H}_0).$$

If we now define the following variables,

$$(3.14) \qquad \boldsymbol{h} = Rj\boldsymbol{H}, \qquad \tau = \omega t, \qquad \boldsymbol{v} = R\omega\boldsymbol{V}, \qquad \gamma = j/\omega,$$

the general linear operator may be written

$$(3.15) \qquad \left\{\nabla^2 - \frac{4}{\lambda^2}\frac{\partial^2}{\partial z^2}\right\}\Phi = 0,$$

with the boundary conditions

$$(3.16) \qquad \{[r(\partial/\partial r) + z(\partial/\partial z)] + 2(k/\lambda) - (4/\lambda^2)z(\partial/\partial z)\}\Phi = 0,$$

where

$$(3.17) \qquad \lambda = \frac{k^2\gamma^2 - \sigma^2}{k\gamma^2 + \sigma}$$

and where $\Phi = -i\sigma(P + \frac{1}{2}H_0^2 + \boldsymbol{H} \cdot \boldsymbol{H}_0)$.

The determination of permissible free hydromagnetic oscillations is now reduced to the solved Poincaré eigenvalue problem. GREENSPAN (1965) has made a comprehensive analysis of the eigenstructure of eqs. (3.15) and (3.16). In Greenspan's and earlier studies, $\lambda$ was the frequency of oscillation, corresponding to $\sigma$ in eq. (3.11). GREENSPAN establishes that $\lambda$ is real and $|\lambda|$ is less than 2. The eigenvalue $\lambda = 0$ is associated with the infinite subset of geostrophic modes of motion. The eigenfunctions are orthogonal polynomials in $r$ and $z$. The problem is as easily solved in an arbitrary spheroid as in a sphere.

In this hydromagnetic case the restrictions on $|\lambda|$ do not restrict $\sigma$, as may be seen in eq. (3.17). Solving that equation for $\sigma$ one obtains

$$(3.18) \qquad \sigma = \tfrac{1}{2}\lambda(-1 \pm [1 + 4\gamma^2 k(k - \lambda)/\lambda^2]^{\frac{1}{2}}) .$$

For $\lambda = 0$ there is the one acceptable solution:

$$(3.19) \qquad \sigma = 0 , \qquad k = 0 ,$$

which represents the subset of magnetogeostrophic flows for all states $\boldsymbol{v} \times \boldsymbol{H_0} = 0$.

A second solution of eq. (3.17) for $\lambda = 0$ is

$$(3.20) \qquad \sigma = \pm \gamma k ,$$

which represents two cylindrical Alfvén waves whose angular phase speed is independent of $r$. Like all the other solutions for $\lambda = 2$, one can show that these Alfvén waves cannot satisfy the boundary conditions of the problem and hence are unacceptable.

An interesting consequence of eq. (3.18) is that unstable waves can occur in the special case $k = 1$. Rewriting eq. (3.18) for $k = 1$ as

$$(3.21) \qquad \sigma = -\tfrac{1}{2}\lambda \pm i([4\gamma^2(\lambda - 1)/\lambda^2] - 1)^{\frac{1}{2}} ,$$

one sees that instability occurs when

$$(3.22) \qquad \gamma^2 > \tfrac{1}{4}\lambda^2/(\lambda - 1) .$$

We establish in the following that there are $k = 1$ modes with $+2 > \lambda \geqslant 1$. Hence the minimum $\gamma^2$ for instability is $\gamma^2 \geqslant 1$.

However, for the geophysical problem in which we are interested, $\gamma^2$ is very small. When this is so, $\sigma$ has the roots

$$(3.23) \qquad \sigma = \gamma^2[(k^2/\lambda) - k] , \qquad\qquad \sigma = -\lambda .$$

As no explicit study of the possible dispersion relations $\lambda = \lambda(k)$, here $\sigma = \sigma(k)$, has yet been made, several will be deduced in the following paragraphs.

The general solution of eq. (3.15) is

$$(3.24) \qquad \Phi_{nk} = P_n^k(\eta) P_n^k(\mu) ,$$

where $P_n^k(x)$ is an associated Legendre polynomial and

$$\gamma = \eta_0^{-1}(1 - \mu^2)^{\frac{1}{2}}(1 - \eta^2)^{\frac{1}{2}}, \qquad z = (1 - \eta_0^2)^{\frac{1}{2}}\mu\eta.$$

The new independent variable $\eta$ should not be confused with the (neglected) magnetic diffusivity in eq. (3.3).

From the boundary conditions the eigenvalues problem may be written

$$(3.25) \qquad (1 - \eta_0^2)[\partial P_n^k(\eta)/\partial \eta]_{\eta=\eta_0} = kP_n^k(\eta_0),$$

where $\eta_0 = \frac{1}{2}\lambda$. As there are many possible roots $\eta_0$ of the polynomial eq. (3.25) for each choice of $k$, $n$, a particular eigenfunction $\Phi_{nk}$ can have many eigenvalues. However, the fields $v$ and $h$ derived from $\Phi_{nk}$ are explicit functions of $\eta_0$, so that each $\eta_0(n, k)$ corresponds to a set of fields $v$, $h$, $P$ which differ from each other.

From the properties of the associated Legendre polynomials one may rewrite eq. (3.25) as:

$$(3.26) \qquad (n\eta_0 + k)P_n^k(\eta_0) = (n + k)P_{n-1}^k(\eta_0).$$

Two simple relations between $\eta_0$ and $k$ can be found from eq. (3.26) for those eigensolutions $\Phi_{nk}$ with linear and quadratic dependence on $z$. The asymptotic values of $\eta_0$ for $n \gg k$ can also be determined. Other relations are immersed in the algebraic complexity of the eigenvalue problem. For the case $n = k + 1$, the associated Legendre polynomials have the property

$$(3.27) \qquad P_{k+1}^k(\eta_0)/P_k^k(\eta_0) = 2k + 1)\eta_0;$$

hence

$$(3.28) \qquad (\eta_0)_{k+1,k} = k/k + 1.$$

The corresponding unnormalized eigenfunction is

$$(3.29) \qquad \Phi_{k+1,k}(r, z) = zr^k.$$

The angular phase velocities of hydromagnetic waves for this class of solutions is found to be:

$$(3.30) \qquad C_p = \sigma/k = \gamma^2[(k/2\eta_0) - 1], \qquad C_p = -2\eta_0/k,$$

$$(3.31) \qquad (C_p)_{k+1,k} = \gamma^2[\tfrac{1}{2}k(k + 1) - 1], \qquad (C_p)_{k+1,k} = -2k/(k + 1).$$

Among the interesting properties of these solutions is that their lowest zonal mode, $k = 1$, has both the «high speed» hydrodynamic solution $C_p = -1$

and the stationary solution $C_p = 0$. This lowest mode represents a small tilt of the $j$-axis away from the $\omega$-axis. The result above indicates that solutions to this problem are not sensitive to such a tilt.

Another property of eq. (3.29) is that it is a solution which would persist in a radially stratified density field, since it represents eastward motions in spherical shells.

The second case which can be treated is $n = k + 2$. The two acceptable roots of eq. (3.26) are

$$(3.32) \qquad (\eta_0)_{k+2} = \frac{1}{k+2}\left[1 \pm \frac{(k+1)(k+3)}{2k+3}\right];$$

hence

$$(3.33) \qquad (C_p)_{k+2,k} = \gamma^2\left(\frac{k(k+2)}{2\{1 \pm [(k+1)(k+3)/(2k+3)]^{\frac{1}{2}}\}} - 1\right) \rightarrow \pm \gamma^2\frac{k^{\frac{3}{2}}}{2^{\frac{1}{2}}}.$$

The corresponding unnormalized eigenfunctions are

$$(3.34) \qquad \Phi_{k+2,k} = r^k[(2k+3)(1-\eta_0)r - 2(k+1)\{1 - (2k+3)\eta_0z^2\}].$$

These oscillations propagate both east and west, and their vertical velocity depends linearly on $z$ throughout the fluid. One also notes that

$$(3.35) \qquad \lambda_{3,1} = -0.1767, \ +1.510.$$

Hence this is the first $k = 1$ mode with a positive $\lambda > 1$; it would be unstable if $\gamma^2 > 1.117$.

We conclude that a class of low-frequency modes moves to the east, another class moves to the west, and that in the many dispersion relations $C_p$ is proportional to $\pm k^s$, where $1 < s < 2$.

In contrast to the conventional Alfvén waves, the linear solutions described in this Section are not nonlinear solutions. Hence, they can contribute to the advection of momentum in their nonlinear form.

Nothing has been found in the linear or nonlinear aspects of the free-wave solutions to suggests a preference for westward motion. Hence, it may be that the apparent westward motion of the earth's field is due to a retrograde rotation of a significant fraction of the core as suggested by BULLARD, FREEDMAN, GELLMAN and NIXON in 1950.

Another possibility is that, owing to the nature of the energy source, the most unstable waves move west. Laboratory studies of the flow induced in rotating spheroids by forced precession indicate just such instability of *quasi*-two-dimensional waves moving to the west. Although the motions in the earth's core seem quite turbulent, the precession of the earth may induce sufficient toroidal flow to selectively excite the westward-moving waves.

BIBLIOGRAPHY

H. P. Greenspan: *Journ. Fluid Mech.*, **22**, 449 (1965).
R. Hide: *Science*, **157**, 55 (1967).
W. V. R. Malkus: *Journ. Fluid Mech.*, **28**, 793 (1967).
H. P. Pao: *Phys. Fluids*, **9**, 1254 (1966).
H. Poincaré: *Bull. Astron.*, **27**, 321 (1910).
E. C. Bullard, C. Freeman, H. Gellman and J. Nixon: *Phil. Trans.*, A **243**, 67 (1950).

## 4. – Flow in a precessing spheroid.

The experimental study of fluid flow in a precessing spheroidal cavity has led to the isolation of four significantly different regimes. First, the tilt of the fluid's axis of rotation away from the axis of rotation of the container was indicative of the Poincaré type of solution. Second, a laminar zonal shearing flow appeared, whose intensity was a function of the Poincaré tilt. Thirdly, when this zonal flow became intense enough, a wavy instability was observed to occur. Lastly, with further precessional torque, the flow became quite turbulent.

The observed tilt in the laminar regime was in agreement with the classical theory described in the previous Section. The emphasis in this Section will be the on laminar zonal flow. I will conclude the Section with a discussion of progress towards an understanding of the wavy instability and some brief comments on the observations of the turbulent regime.

The origin of the zonal shearing flow must be the failure of the Poincaré solution to exactly satisfy the viscous boundary conditions. Roberts and Stewartson studied the linear boundary layer phenomenon. They concluded that the addition of a thin boundary layer flow met the viscosity requirements, and left Poincaré's solution for the fluid motion essentially unaltered. However, in contrast to their conclusion, the experiments suggest that the intensity of the observed zonal flow becomes greater as the viscosity of the fluid is decreased. It was clear that finite-amplitude precessional flow in the limit of vanishing viscosity would be very different from the classical flow found by assuming zero viscosity initially. I traced the origin of these zonal flows to the non-linear momentum advection in the thin boundary layers. With the assistance of my colleague Busse, a first interpretation of the internal flow linked it with the resonant excitation of internal free modes driven by the boundary layer stresses. A detailed theoretical treatment of the problem in the limit of vanishingly small viscosity and small, but finite, amplitude has been prepared by Busse (1968). It is this work which will be discussed in the following.

The basic equations appropriate for steady flows in the precessing system are: first for the continuity of mass

(4.1)                                   $\nabla \cdot v = 0$ ,

second for the continuity of momentum

(4.2)                         $- E\nabla^2 v + 2\Omega \cdot v + v \cdot \nabla v + \nabla P = 0$ ,

here scaled so that the velocity on the surface $S$ of the container is written as:

(4.3)                                   $v = k \times r$ ,

where $k$ is the unit normal vector along the axis of rotation, and $r$ is the position vector measured from the center of the system. The scaling is also such that the viscous term appear multiplied by the Ekman number

(4.4)                                   $E = \dfrac{\nu}{\omega R^2}$ ,

where $\nu$ is the kinematic viscosity, $\omega$ the angular velocity of rotation and $R$ the principal radius of the container. The shape of the container is given by:

(4.5)                                   $r^2 + \eta(k \cdot r)^2 = 1$ ,

where $\eta$ is the ellipticity of the oblate spheroid.

The boundary layer theory will be used to determine the flow to second order in amplitude. (A good reference for this use of singular perturbation theory is the book « The Theory of Rotating Fluids » by GREENSPAN.) We first express the velocity of the fluid as the sum of an interior inviscid velocity plus a boundary layer contribution as

(4.6)                                   $v = u + \tilde{u}$ ,

where $u$ is the (inviscid) interior flow and $\tilde{u}$ is the boundary layer flow, presumed to vanish in the interior. The sum of these two velocities is to satisfy the correct viscous boundary condition. The spatial variable normal to the boundary is scaled in terms of

(4.7)                                   $\xi \equiv - [r(S) - r] \cdot n E^{-\frac{1}{2}}$ ,

where $n$ is the unit vector normal to the surface $S$.

From eqs. (4.7), (4.6), and (4.1), one may write

$$(4.8) \qquad \nabla \cdot \tilde{\boldsymbol{u}} = - E^{-\frac{1}{2}} \frac{\partial}{\partial \xi} \, \tilde{\boldsymbol{u}} \cdot \boldsymbol{n} + \boldsymbol{n} \cdot \nabla \times (\boldsymbol{n} \times \tilde{\boldsymbol{u}}) + \boldsymbol{n} \cdot \tilde{\boldsymbol{u}} \nabla \cdot \boldsymbol{n} = 0 \, .$$

The magnitude of the departure of total velocity from solid rotation is defined as $\varepsilon$, the amplitude of the relative motion. The boundary layer and interior velocities are expanded in a double series in $E^{\frac{1}{2}}$ and the amplitude $\varepsilon$ as

$$(4.9) \qquad \left\{ \begin{array}{ll} \boldsymbol{u} = \sum_{\substack{\mu=0 \\ \nu=0}}^{\infty} \varepsilon^{\mu} E^{\nu/2} \boldsymbol{u}_{\nu}^{\mu} \, , & \boldsymbol{u}_0^0 \equiv \boldsymbol{k} \times \boldsymbol{r} \, , \\[18pt] \tilde{\boldsymbol{u}} = \sum_{\substack{\mu=0 \\ \nu=0}}^{\infty} \varepsilon^{\mu} E^{\nu/2} \tilde{\boldsymbol{u}}_{\nu}^{\mu} \, , & \tilde{\boldsymbol{u}}_0^0 \equiv 0 \, . \end{array} \right.$$

The basic state is chosen as the one of solid rotation. The inhomogeneous forcing due to the precession will be introduced in the $\varepsilon E^{\frac{1}{2}}$ order to remove the indeterminancy of the problem. We treat $\eta$, the ellipticity, and $E^{\frac{1}{2}}$ as of the same order of magnitude to avoid a triple expansion.

The formal ordering of the sequence of problems is as follows

$$(4.10) \qquad \left\{ \begin{array}{l} \nabla \cdot \boldsymbol{u}_0^1 = 0 \, , \qquad \boldsymbol{u}_0^1 \cdot \boldsymbol{n} = 0 \ \text{ on } \ S \, , \\[8pt] \boldsymbol{u}_0^0 \cdot \nabla \boldsymbol{u}_0^1 + \boldsymbol{u}_0^1 \cdot \nabla \boldsymbol{u}_0^0 = - \nabla P_0^1 \, , \end{array} \right.$$

where our purpose will be to find a class of $\boldsymbol{u}_0^1$ fields compatible with these equations. Secondly,

$$(4.11) \qquad \left\{ \begin{array}{l} \tilde{\boldsymbol{u}}_0^1 + \boldsymbol{u}_0^1 = 0 \ \text{ on } \ S \, , \\[8pt] E^{-\frac{1}{2}} \tilde{\boldsymbol{u}}_0^1 \cdot \boldsymbol{n} = 0 = \tilde{P}_0^1 \, , \\[8pt] 2 \boldsymbol{\Omega} \times \tilde{\boldsymbol{u}}_0^1 + \boldsymbol{u}_0^0 \cdot \nabla \tilde{\boldsymbol{u}}_0^1 + \tilde{\boldsymbol{u}}_0^1 \cdot \nabla \boldsymbol{u}_0^0 = \boldsymbol{n} \, \dfrac{\partial \tilde{P}_1^1}{\partial \xi} + \dfrac{\partial^2}{\partial \xi^2} \, \tilde{\boldsymbol{u}}_0^1 \, . \end{array} \right.$$

Where $P_0^1$ and $P_1^1$ are the pressure terms in the expansion as in eq. (4.9). Our task is here to find $\tilde{\boldsymbol{u}}_0^1$ in terms of the $\boldsymbol{u}_0^1$ found in eq. (4.10) above. Thirdly,

$$(4.12) \qquad - \frac{\partial \tilde{\boldsymbol{u}}_1^1 \cdot \boldsymbol{n}}{\partial \xi} + \boldsymbol{n} \cdot \nabla \times (\boldsymbol{n} \times \tilde{\boldsymbol{u}}_0^1) = 0 \, .$$

Here we are to find $\tilde{\boldsymbol{u}}_1^1$ in terms of $\tilde{\boldsymbol{u}}_0^1$, and through the boundary condition to determine $\boldsymbol{u}_1^1 \cdot \boldsymbol{n}$ on the surface $S$. Fourthly,

(4.13)
$$\begin{cases} \nabla \cdot \boldsymbol{u}_1^1 = 0 \,, \\ 2\boldsymbol{\Omega} \times \boldsymbol{u}_1^1 + \boldsymbol{u}_0^0 \cdot \nabla \boldsymbol{u}_1^1 + \boldsymbol{u}_1^1 \cdot \nabla \boldsymbol{u}_0^0 + \nabla P_1^1 = + \left( \dfrac{1}{\varepsilon E^{\frac{1}{2}}} \right) \boldsymbol{\Omega} \times \boldsymbol{u}_0^0 \,. \end{cases}$$

It is not necessary to find $\boldsymbol{u}_1^1$ itself, however, the solvability condition for $\boldsymbol{u}_1^1$ determines the $\boldsymbol{u}_0^1$ which will be produced by the precessional forcing function.

We anticipate that

$$\varepsilon \boldsymbol{u}_0^1 = (\boldsymbol{\omega} - \boldsymbol{k}) \times \boldsymbol{r} \,,$$

where $\boldsymbol{\omega}$ is a constant vector, will fulfil the solvability condition. One determines from the above sequence of problems that

$$\omega^2 = \boldsymbol{k} \cdot \boldsymbol{\omega} \,, \qquad \varepsilon^2 = 1 - \omega^2$$

and

(4.14)
$$\frac{\boldsymbol{\omega}}{\omega^2} = \boldsymbol{k} + \frac{A\boldsymbol{k} \times (\boldsymbol{\Omega} \times \boldsymbol{k}) + B(\boldsymbol{k} \times \boldsymbol{\Omega})}{A^2 + B^2} \,,$$

where

$$A = \left( 0.259 \sqrt{\frac{E}{\omega}} + \eta \omega^2 + \boldsymbol{\Omega} \cdot \boldsymbol{k} \right), \qquad B = 2.62 \sqrt{E\omega} \,.$$

Equation (4.14) is valid to second order in $\varepsilon$ for the determination of the direction of $\boldsymbol{\omega}$ due to the precession. However, second order terms do cause a change in the zonal flow relative to the vector $\boldsymbol{\omega}$. Second-order theory exactly parallels the linear theory above. One seeks that $\boldsymbol{u}_0^2$ which permits solvability of the equation for $\boldsymbol{u}_1^2$. The boundary conditions for $\boldsymbol{u}_1^2$ are dominated by the boundary layer efflux due to $\tilde{\boldsymbol{u}}_0^1 \cdot \nabla \tilde{\boldsymbol{u}}_0^1$. One may write

(4.15)
$$\boldsymbol{u}_0^2 = \boldsymbol{\omega} \times \boldsymbol{r} f \left( \frac{|\boldsymbol{\omega} \times \boldsymbol{r}|}{\omega} \right),$$

where, with perseverance and skill, Busse has found the $f$ given in Fig. 5.

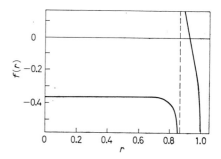

Fig. 5. – Departure from solid rotation in a precessing sheroid, eq. (15).

The infinite shear layer in Fig. 5 lies on the cylinder which intercepts the surface $S$ at $\pm 30°$ latitude relative to the equator determined by $\boldsymbol{\omega}$.

Although interior viscous effects certainly will smear out this intense shear, it is clear that inviscid theory differs dramatically from viscous theory in the limit of vanishing viscosity.

A work which parallels this study was done by doctoral candidate S. Suess on flow due to a tidal bulge. The off-equator component of the bulge gives rise to a $u_0^2$ flow just as in Fig. 5, while the equatorial component of the bulge gives rise to intense shear layers on the axis of the flow. This result is exhibited in Fig. 6.

A representative sample of data for precessing flow in the laboratory exhibits the dramatic features of both Fig. 5 and 6, but strongly smoothed by internal viscous effects. This is

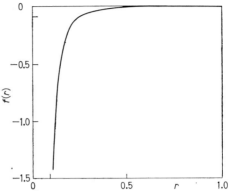

Fig. 6. – Departure from solid rotation in a tidally deformed spheroid (after SUESS).

seen in Fig. 7, which is a tracing from the photographic records of a dye streak introduced into a laminar flow such as we saw in the photographs in the previous Section. The dye was produced electrically in a dilute solution of thymol blue, by means of a straight wire probe extending from the equator of the spheroid to the center of the fluid. The probe was withdrawn and a very slow precession was started. Photographs of the developing flow were taken at 1 minute intervals. A steady toroidal velocity relative to the container was reached after several minutes. The Ekman number for this flow is

$$E = 3.6 \cdot 10^{-6}.$$

In Poincaré-like flows, and in the flows predicted by STEWARTSON and ROBERTS, the dye line of Fig. 7 would not have moved at all. The interior flow in Poincaré core fluid is a smooth zero average diurnal oscillation. However, due to the second-order features caused by the boundary layer efflux, one sees the beginnings of a sharp jet at 30° latitude and considerable shear along the axis of the flow. The jets became much sharper at smaller $E$. Data taken at Ekman numbers of $3 \cdot 10^{-7}$ exhibit pro-grade jets 5 times as intense as shown in Fig. 7.

These shearing flows are observed to become unstable at values of $\varepsilon^2 \geqslant$ $\geqslant 5E^{\frac{1}{4}}F(E)$, where $F(E) \sim 0.4$ in these laboratory observations. The trend of the laboratory observations suggests that $F(E)$ may vary as $E^{-\frac{2}{5}}$ for extremely small $E$. However, this point has negligible theoretical foundation and only a few supporting experimental data. In any instance, if this scaling from the laboratory is appropriate to the Earth, flow in the Earth's core should be very unstable.

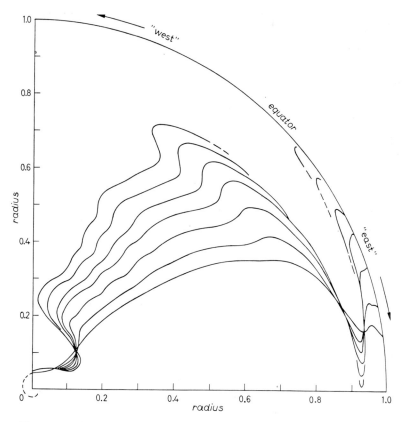

Fig. 7. – Repeated photographs of a dye line in the equatorial plane of a laboratory
precessing flow.

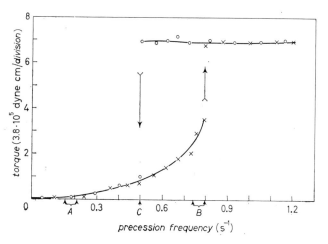

Fig. 8 – Motor torque due to forced precession.

The character of unstable precessional flow has not been studied in any detail. Figure 8 exhibits the one kind of quantitative data yet obtained, namely the torque needed to maintain the flow. This particular value was taken from a spheroid rotating at 900 revolutions per minute. The angle between the rotation precession axis was 90°. The laminar regime of flow extends only to the range $A$), indicated in the figure. While between $A$) and $B$), a wavy nonlinear flow persisted, becoming more and more intense. At $B$) a dramatic transition occurred to a turbulence state which appeared « saturated », since the torque required to maintain it was independent of precession frequency over a wide range. However, on reducing the precession frequency to the point labelled $C$ in Fig. 8, the flow relaxed back to the wavy regime.

Torque records containing hysteresis are commonplace in our studies of the flow due to precession. We anticipate that a transition to hydromagnetic turbulence will exhibit the same finite amplitude behaviour.

## BIBLIOGRAPHY OF SECTION 4

F. H. BUSSE: *Journ. Fluid Mech.*, **33**, 739 (1968).

H. P. GREENSPAN: *Journ. Fluid Mech.*, **22**, 449 (1965).

W. V. R. MALKUS: *Science*, **160**, 259 (1968).

P. H. ROBERTS and K. STEWARTSON: *Astrophys. Journ.*, **137**, 777 (1963).

S. SUESS: Doctoral Dissertation, Department of Planetary Science U.C.L.A. (1969).

# Physical Properties of the Earth's Mantle and Core.

J. A. JACOBS

*The University of Alberta - Edmonton*

## 1. – The variation of density within the Earth.

The density $\varrho$ will depend on the pressure $p$, the temperature $T$, and an indefinite number of parameters $n_i$ specifying the chemical composition *i.e.*

$$(1) \qquad \varrho = \varrho(p,\, T,\, n_i)\,.$$

If $m$ is the mass of material within a sphere of radius $r$, then, since the stress in the Earth's interior is essentially equivalent to a hydrostatic pressure,

$$(2) \qquad \frac{\mathrm{d}p}{\mathrm{d}r} = -g\varrho\,,$$

where

$$(3) \qquad g = \frac{GM}{r^2}$$

and $G$ is the constant of gravitation. The assumption of hydrostatic pressure would be a poor approximation for the stress in the crust but is unlikely to be much in error in the mantle and core. If we consider for the moment a chemically homogeneous layer in which the temperature variation is adiabatic, it follows from equation (1) that

$$(4) \qquad \frac{\mathrm{d}\varrho}{\mathrm{d}r} = \frac{\mathrm{d}\varrho}{\mathrm{d}p}\frac{\mathrm{d}p}{\mathrm{d}r} = \frac{-g\varrho^2}{k_s}\,,$$

where $k_s$ is the adiabatic incompressibility defined by the equation:

$$(5) \qquad \frac{1}{k_s} = \frac{1}{\varrho}\left(\frac{\partial \varrho}{\partial p}\right)_s$$

and $S$ is the entropy. A homogeneous region is here defined as one in which there are no significant changes of either phase or chemical composition. Combining equations (3) and (4), we have

$$(6) \qquad \frac{\mathrm{d}\varrho}{\mathrm{d}r} = \frac{-Gm\varrho^2}{k_s r^2} \,.$$

If $V_p$ and $V_s$ are the velocities of $P$ and $S$ waves, then

$$(7) \qquad V_p = \sqrt{\frac{k_s + \frac{4}{3}\mu}{\varrho}}$$

and

$$(8) \qquad V_s = \sqrt{\frac{\mu}{\varrho}} \,,$$

where $\mu$ is the rigidity. Thus

$$(9) \qquad \varphi = \frac{k_s}{\varrho} = V_p^2 - \frac{4}{3} V_s^2 \,.$$

The distribution of $V_p$ and $V_s$ and hence $\varphi$ throughout the Earth can be determined from seismic travel time curves if $V_p$ and $V_s$ are monotonic functions of $r$. The radial distribution of these velocities is critical for a determination of the density. The subject is in a state of rapid development at the moment —especially the question of lateral variations in the upper mantle.

Since $\mathrm{d}m/\mathrm{d}r = 4\pi\varrho r^2$, a second order differential equation for $\varrho = \varrho(r)$ can be written down by differentiating equation (6),

$$(10) \qquad \frac{\mathrm{d}^2\varrho}{\mathrm{d}r^2} - \frac{1}{\varrho}\left(\frac{\mathrm{d}\varrho}{\mathrm{d}r}\right)^2 + P(r)\frac{\mathrm{d}\varrho}{\mathrm{d}r} + Q(r)\varrho^2 = 0 \,,$$

where

$$(11) \qquad P(r) = \frac{2}{r} + \frac{1}{\varphi}\frac{\mathrm{d}\varphi}{\mathrm{d}r} \quad \text{and} \quad Q(r) = \frac{4\pi G}{\varphi} \,.$$

Equation (10), which was first obtained by ADAMS and WILLIAMSON in 1923, may be integrated numerically to obtain the density distribution in those regions of the Earth where chemical and nonadiabatic temperature variations may be neglected.

Figure 1 shows the gross features of the velocity depth curves in the mantle and core according to JEFFREYS and GUTENBERG. During 1940-1942, BULLEN divided the Earth into a number of regions based on velocity-depth curves. His nomenclature has since been widely used, and, in spite of uncertainties

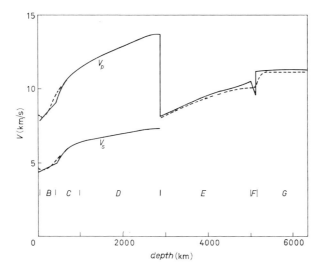

Fig. 1. – Seismic velocities of $P$ and $S$ waves as a function of depth according to Jeffreys and Gutenberg. ——— JEFFREYS. – – – GUTENBERG.

in the boundaries between the different regions, continues to serve as a useful basis in discussing the Earth's interior. The upper mantle consists of region $B$ extending from the base of the crust (region $A$) to a depth of about 400 km, and region $C$ which is a transition zone between depths of about 400 and 1000 km. The lower mantle, below a depth of about 1000 km, is called region $D$. BULLEN later subdivided region $D$ into $D'$ and $D''$—the region $D''$ being the bottom 200 km. The core is divided into an outer core $E$, a transition region $F$ and an inner core $G$. The outer core $E$ is fluid and the inner core $G$ most probably solid. During recent years much finer detail has been found, particularly in the upper mantle (regions $B$ and $C$) and in the vicinity of the inner core (region $F$). This will be discussed later.

Any density distribution must satisfy two conditions—it must yield the correct total mass of the Earth and the correct moment of inertia about its rotational axis. Using these two conditions and a value $\varrho_1$ of 3.32 g(cm)$^{-3}$ for $\varrho$ at the top of layer $B$ of the mantle (assumed to be at a depth of 33 km), BULLEN applied eq. (10) throughout the regions $B$, $C$, and $D$. He then found that this led to a value of the moment of inertia $I_c$ of the core greater than that of a uniform sphere of the same size and mass. This would entail the density to decrease with depth in the core and would be an unstable state in a fluid.

If the temperature gradient is not adiabatic, BIRCH [1] (*) showed that

_____

(*) See Appendix A.

the right-hand side of eq. (4) may be modified by the inclusion of an additional term, *viz.*

$$(12) \qquad\qquad \frac{d\varrho}{dr} = -g\varrho/\varphi + \alpha\varrho\tau ,$$

where $\alpha$ is the volume coefficient of thermal expansion and $\tau$ the difference between the actual temperature gradient and the adiabatic gradient. Allowance for the second term in eq. (12) increased the value of $I_c$ still more. A reasonable value for $I_c$ could be obtained by increasing the initial value $\varrho_1$, but only if an impossibly high value (at least 3.7 g(cm)$^{-3}$) was chosen. Thus the assumption of chemical homogeneity must be in error; the region where this assumption is most likely to be invalid is region $C$ where there are large changes in the slope of the velocity depth curves. In his original Earth model $A$, BULLEN thus used the Adams-Williamson eq. (10) in regions $B$ and $D$ while in region $C$ he fitted a quadratic expression in $r$ for $\varrho = \varrho(r)$.

In the outer core (region $E$) eq. (10) is likely to apply, and values of $\varrho$ down to a depth of about 5000 km can be obtained with some confidence. In the core one boundary condition is $m = 0$ at $r = 0$ but lack of evidence on the value of the density $\varrho_0$ at the centre of the Earth leads to some indeterminacy in the density distribution in regions $F$ and $G$. However since these regions constitute only about 1 per cent of the Earth's total volume the density distribution within $E$ can be estimated fairly precisely.

BULLEN (*) also found that strong controls on permissible density values are exercized by various moment of inertia criteria. In particular he showed that the minimum possible value of $\varrho_0$ is 12.3 g (cm)$^{-3}$ and that increasing the value of $\varrho_0$ by 5 g (cm)$^{-3}$ affects the formally computed densities elsewhere by maximum amounts of only 0.03 g (cm)$^{-3}$ in the mantle and 0.4 g (cm)$^{-3}$ in the outer core. BULLEN derived density distributions on two fairly extreme hypotheses, i) $\varrho_0 = 12.3$ g (cm)$^{-3}$ and ii) $\varrho_0 = 22.3$ g (cm)$^{-3}$ (this value being taken quite arbitrarily). A model with density values midway between those of these two hypotheses has been called model $A$. More recent evidence indicates that $\varrho_0$ is probably much nearer its minimum value and that a model based on $\varrho_0 = 12.3$ g (cm)$^{-3}$ (model $A$-i) is more likely to be correct. There have been a number of more recent determinations of the density distribution within the Earth using additional data provided by the free vibration of the Earth excited by the Chilean and Alaskan earthquakes of May 1960 and March 1964. There have also been new approaches to the problem. These will be discussed after the next Section which shows how other physical parameters of the Earth may be determined once the density distribution is known.

---

(*) A good account of this early work can be found in BULLEN [2].

## 2. – Pressure distribution, variation of acceleration due to gravity, and elastic constants within the Earth.

From eqs. (2) and (3), it follows that

$$(13) \qquad \frac{\mathrm{d}p}{\mathrm{d}r} = \frac{- Gm\varrho}{r^2}.$$

Hence by numerical integration, the pressure distribution may be obtained once the density distribution has been determined. Since the density is used only to determine the pressure gradient, the pressure distribution is insensitive to small changes in the density distribution and may be determined quite accurately. The uncertainty in the values for the regions $F$ and $G$ is probably of the order of 3 per cent and for all other regions less than 1 per cent.

The variation of $g$ can be calculated from eq. (3); its value does not differ by more than 1 per cent from the value 990 cm (sec)$^{-2}$ until a depth of over 2400 km is reached. On the other hand, the values of $g$ deep within the Earth are sensitive to changes in density and values below 4000 km may be in error by as much as 5 per cent.

From a knowledge of the density distribution, it is easy to compute values of the elastic constants. Thus eqs. (8) and (9) give $\mu$ and $k$ directly. From the known relationships between the elastic constants, it is possible to compute the distribution of Young's modulus $E$ and Poisson's ratio $\sigma$. In particular

$$(14) \qquad \sigma = \frac{3k - 2\mu}{6k + 2\mu} = \frac{V_p^2 - 2V_s^2}{2(V_p^2 - V_s^2)}$$

and is thus independent of the density $\varrho$.

## 3. – Density distribution within the Earth — more recent determinations.

Additional information on the physical properties of the Earth's mantle and core has been provided by an analysis of the free vibrations of the Earth excited by the great Chilean earthquake of 22 May, 1960. Both spheroidal and toroidal oscillations were observed; however neither the Jeffreys nor the Gutenberg velocity distributions combined with Bullen's density distribution are consistent with the longer period free oscillation data. The higher order free oscillations also provide some of the best evidence for the existence of a low velocity zone in the upper mantle first advocated by Gutenberg.

LANDISMAN et al. [3] considered the inverse problem—that of determining the radial distribution of $p$ $V_p$ and $V_s$ from the travel time curves for $P$ and $S$

waves and the free periods of the Earth. Their method is independent of the assumptions of homogeneity and an adiabatic temperature gradient (except in the region of the outer core). They investigated a number of earth models subject to the constraints fixed by a density of 3.32 g (cm)$^{-3}$ at the top of the mantle, the total mass of the Earth, and its moment of inertia about its axis of rotation. The outer core was taken to be chemically homogeneous with an adiabatic temperature gradient and in it the Adams-Williamson equation was used to determine the density. In their model $M_1$, the lower core was considered to be chemically inhomogeneous following the model proposed by BULLEN [4, 5] and BOLT [6]. In their model $M_3$, the entire core was taken to be a homogeneous adiabatic fluid. As with all models, Hooke's law was assumed to be valid throughout the Earth and no density inversions were permitted at any depth.

The density in both their models $M_1$ and $M_3$ is about 10 per cent lower near the base of the mantle than that predicted by the Bullen $A$-i distribution. Constant densities were also found for both these models at depths between about 1600 and 2800 km. A super-adiabatic temperature gradient of $(4 \div 5)$ °C/km would be required to explain such a result—this would lead to excessive temperatures at the core-mantle boundary. If the vanishing density gradient were the result of a compositional change such as the depletion of iron, a concomitant increase of almost 10 per cent would be expected in the shear velocity gradient. A super-adiabatic temperature gradient of about 2 °C/km combined with a depletion of iron is a possible explanation. From depths of 500 to 1100 km, densities for models $M_1$ and $M_3$ are slightly lower and from 1100 to 1800 km slightly higher than those of the Bullen $A$-i distribution. For model $M_1$, the density of the outer core is found to be about 2 per cent higher than that for Bullen $A$-i), while for model $M_3$ it is about 4 per cent higher. The central density for model $M_1$ is about 15.42 g (cm)$^{-3}$ and that for model $M_3$ about 12.63 g (cm)$^{-3}$.

PEKERIS [7] has also obtained density distributions in the Earth using the observed periods of the free oscillations of the Earth, without using the Adams-Williamson assumption of homogeneity and adiabaticity for any region (including the outer core). The density distribution $\varrho(r)$ was represented by 50 pivotal values $\varrho(r_k)$ with linear variations in between and with discontinuities at the Moho (the crust-mantle interface) and at the core-mantle boundary. The $\varrho_k$ were varied by the method of steepest descent so as to minimize the sum of the squares of the residuals of all the observed periods. PEKERIS found that a somewhat better fit to the observed spectral data is obtained for a Gutenberg-type velocity distribution (with a low-velocity layer) than for the Jeffreys distribution. The final density distributions for the various models in the depth range $(400 \div 1200)$ km converge in a narrow band close to that of Bullen's model $A$-i. In the first 400 km, the density is nearly constant for Gutenberg-type models with a tendency towards negative density gradients in the first 200 km. All models give a density at the outer boundary of the core close to 10.0 g (cm)$^{-3}$; also as is to be expected, the density distribution in the inner core has little effect on the spectrum as a whole. The same degree of improvement in the spectral fit obtained by varying the density alone can also be achieved by varying the shear velocity alone. The new shear velocity distribution shows a tendency to be constant or minimum in the depth range $(100 \div 200)$ km.

BIRCH [8] has estimated the density distribution in the Earth based in part on the empirical observation that for silicates and oxides of about the same iron content, there is an approximate linear relationship between the density $\varrho$ and the velocity $V_p$ of compressional waves, i.e.

$$(15) \qquad\qquad \varrho = a + bV_p .$$

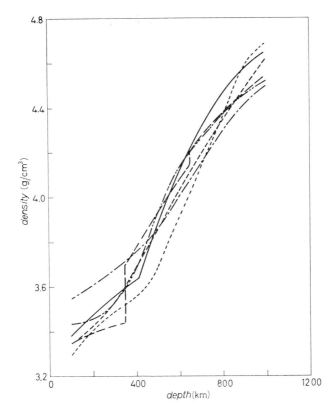

Fig. 2a). – Density distribution within the Earth according to various authors. —— BULLEN A-i. $\cdots$ CLARK and RINGWOOD (pyrolite). —·—·— CLARK and RINGWOOD (eclogite). – – – LANDISMAN et al. M.I. —··—··— Birch solution 1. — — — HADDON and BULLEN HB$_1$.

In the lower mantle and core BIRCH determined the density distribution from the Adams-Williamson equation and only used the empirical relation (15) in the upper mantle and transition zone, *i.e.*, the Adams-Williamson equation is used where the change of density is most probably determined by compression alone, and the empirical rule for the upper mantle, where there are high thermal gradients, and the transition region where there are phase changes. Two models were considered. In the first (Solution I) the constant $b$ was given the value 0.328 $(g (cm)^{-3})/(km (sec)^{-1})$ as found for rocks and crystals of low mean atomic weight. The constant $a$ and the mass of the core are then the only adjustable parameters and are determined by the total mass and moment of inertia of the Earth. For the second model (Solution II) BIRCH chose the constant $a$ so that the density at 33 km is 3.32 $g (cm)^{-3}$ and the adjustable parameters are then $b$ and the mass of the core. The density in the lower mantle is about 1 per cent higher and in the core about 1 per cent lower in the second than in the first model, and are very similar to Bullen's model A-i. The densities in the lower mantle are in good agreement with shock wave measurements on rocks having FeO contents in the range $(10 \pm 2)$ per cent by

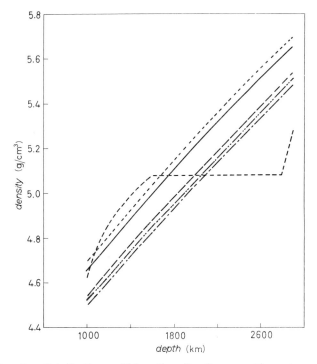

Fig. 2*b*). – Density distribution within the Earth according to various authors ——— BULLEN A-i. ··· CLARK and RINGWOOD (pyrolite). —·—·— CLARK and RINGWOOD (eclogite). – – – LANDISMAN *et al.* M.I. —··—··— Birch solution 1. — — — HADDON and BULLEN HB₁:

weight and may be accounted for in terms of mixtures of close-packed oxides, silica transforming to stishovite in the transition layer. A constant density in the lower mantle, combined with rising velocities, as suggested by LANDISMAN *et al.* [3] is not compatible with Birch's results. However the value of the central density for Landisman's model $M_3$ is almost identical with that of Birch's models.

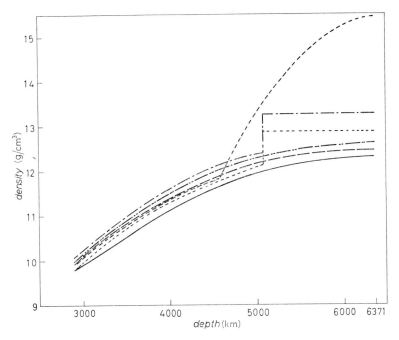

Fig. 2c). – Density distribution within the Earth according to various authors. ——— BULLEN A-i.  · · · CLARK and RINGWOOD (pyrolite). —·—·— CLARK and RINGWOOD (eclogite). – – – LANDISMAN *et al.* M.I.  —··—··— Birch solution 1.  — — — HADDON and BULLEN HB$_1$.

CLARK and RINGWOOD [9] have estimated densities in the Earth using petrological models of the upper mantle constructed on the assumption of an overall pyrolite (ultrabasic) composition and an eclogite composition. There is an almost exact agreement between their pyrolite model and Bullen's model A-i in the lower mantle. Their use of petrological arguments in the upper mantle and transition region, however, lead to densities in these regions which are systematically lower than Bullen's values. Figure 2 a), b), c) shows the density distribution for some of the more recently proposed Earth models together with Bullen's model A-i for comparison. The HB$_1$ model (HADDON and BULLEN [10]) is discussed later in this Section.

PRESS [11, 12] has used a Monte Carlo inversion method to obtain a number of Earth models using as data 97 eigenperiods, travel times of $P$ and $S$ waves and the mass and moment of inertia of the Earth. The Monte Carlo method uses random selection to generate large numbers of models in a computer, subjecting each model to a test against geophysical data. Only those models whose properties fit the data within prescribed limits are retained. This procedure has the advantage of finding models without bias from preconceived or oversimplified notions of Earth structure. Monte Carlo methods also offer the advantage of exploring the range of possible solutions and indicate the degree of uniqueness obtainable with currently available geophysical data.

PRESS was able later [13] to speed up considerably his Monte Carlo procedures. Using new, more extensive and more accurate data he was able to

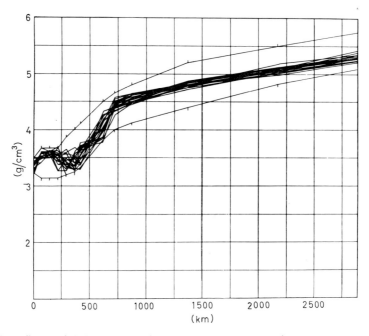

Fig. 3. – Successful density distributions in the mantle (after F. PRESS [13]).

find a larger number of successful models. Figures 3 and 4 show 27 successful density distributions in the mantle and core respectively. Of the millions of Earth models generated and examined, every successful model showed a low-velocity zone for shear waves in the sub-oceanic mantle centered at depths between 150 and 250 km. Again, although density values in the mantle just below the Moho cover the entire permissible range, in the vicinity of 100 km all values fell in the narrow band $(3.5 \div 3.6)$ g(cm)$^{-3}$. This value of the density of the lithosphere near 100 km is so high as to narrow the range of its possible composition to an eclogitic facies.

Large density and velocity gradients were found in the transition zone in the mantle without prior assumption of an equation of state. In particular high-density gradients are localized near $(350 \div 450)$ km and $(550 \div 800)$ km, the details differing between models. These are interpreted as phase transitions and will be discussed later in Sect. 6. Upper mantle models which are

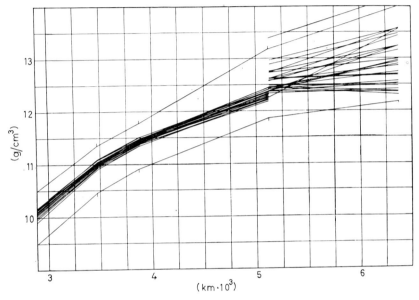

Fig. 4. – Successful density distributions in the core (after F. PRESS [13]).

more complex, with large fluctuations in the shear velocity and density, fit better than « standard » models. Such complexity might be expected if the mantle is chemically and mineralogically zoned and if high-thermal gradients and partial melting take place. The magnitude of the fluctuations suggests that the zoning is lateral and that the mantle is variable in composition laterally ranging from pyrolite to eclogite. PRESS also confirmed that the Earth's core is inhomogeneous. The density at the top of the core is constrained to the narrow range $(9.9 \div 10.2)$ $g(cm)^{-3}$, a value appropriate for a mixture of iron with about 15 wt per cent silicon. Changes in the currently accepted value of the core radius range from $-3$ to $+10$ km.

HADDON and BULLEN [10] have constructed a series of new Earth models (HB) incorporating free oscillation data, consisting of the observed periods of fundamental spheroidal and toroidal oscillations for $0 \leqslant n \leqslant 48$ and $2 \leqslant n \leqslant 44$ respectively and of first and second spheroidal overtones for $n \leqslant 20$. The data were taken from records of both the Chilean and Alaskan earthquakes of May 1960 and March 1964. Their procedure was to start from models derived independently of the oscillation data and to produce a sequence of models showing improved agreement with all the available data. In passing from one model to its successor a guiding principle was to introduce and vary one or more parameters in the model description at any stage to satisfy the oscillation data. They thus try to establish models described in terms of the minimum number of parameters demanded by the data. Thus a major difference in principle between their method and Press' [11-13] Monte Carlo procedure is the comparatively large number of parameters that PRESS has permitted to be randomly varied. HADDON and BULLEN point out that the predominance of complex models found by Press is inherent in his method and that a simple random walk would automatically have a low probability. Press' results however broadly confirm Haddon and Bullen's findings. HADDON and BULLEN also point out that the « average » Earth to which average periods of free oscillation modes relate is not necessarily the same as an Earth model to which the currently available average seismic body wave travel times apply, earthquake epicentres and recording stations not being randomly distributed over the Earth's surface.

One of the intentions of Haddon and Bullen was to examine the conclusion of Landisman *et al.* [3] that the oscillation data require $d\varrho/dr$ to be abnormally low ($\simeq 0$) throughout much of the lower mantle. They can accept this conclusion only if the core radius is kept unchanged: however an increase in the core radius of some ($15\div20$) km not only satisfies the oscillation data but also leads to reduced density gradients in region $B$ and normal gradients throughout most of the lower mantle (region $D$). Additional evidence indicates that the effect of damping on the oscillation periods may be significant —if so the increase needed in the core radius may be somewhat less than 15 km. This evidence would allow some reduction in the density gradient in the lower mantle but not nearly so much as that demanded by LANDISMAN *et al.* Another result of Haddon and Bullen's work is a reduction in the thickness of the crustal layer $A$ from 33 to 15 km. This is a result of the changes made in $V_s$ and $\varrho$ in region $B$ in the course of fitting the oscillation data. Such a reduction is reasonable since it involves a change to a crust that may be more representative of average conditions in the Earth.

It should be emphasized that the overall density distribution is not drastically changed by taking into account the combined effects of the revised moment of inertia of the Earth (as determined by satellite data) and the observational data on the free vibrations of the Earth. Inside the mantle the largest differences in $\varrho$ between models A-i and $HB_1$ is only 0.15 g(cm)$^{-3}$, and inside the core the values of $\varrho$ in model $HB_1$ exceed those in Model A-i at all levels by amounts between 0.2 and 0.3 g(cm)$^{-3}$.

## 4. – Bullen's compressibility-pressure hypothesis (*).

From the results of his Earth model A BULLEN found that there was no noticeable difference in the gradient of the incompressibility $dk/dp$ between the base of the mantle and the top of the core. Moreover there was only a 5 per cent difference in the value of $k$ across the core-mantle boundary. These features are in marked contrast to the large changes in the density and rigidity at the boundary. The change in $k$ is a diminution from the mantle to the core. However interpolation between experimental data at $10^5$ atmospheres and theoretical studies at $10^7$ atmospheres, indicates that in the transition from the mantle to the core, a slight increase in $k$ could be expected for materials likely to occur in the Earth's interior. Because of the smallness in the change in $k$ across the core-mantle boundary and because this change is opposite in sign to that predicted by such an interpolation, BULLEN [14, 15] proposed another Earth model B in which he assumed that $k$ and $dk/dp$ are smoothly varying functions throughout the Earth below a depth of about 1000 km. This hypothesis, called the compressibility pressure $(k, p)$ hypothesis, implies that, at high pressures, the compressibility of a substance is independent of its chemical composition. More recent evidence indicates that the hypothesis as stated is a little too general and that there is some small but significant dependence of $k$ on the representative atomic number as well as on $p$.

On the basis of his hypothesis, BULLEN found that there must be a concentration of more dense material near the base of the mantle (region $D''$). This material could be a mixture of metallic iron with silicates near the core boundary or an iron sulfide phase at the base of the mantle.

If the entire core is liquid so that $V_s = 0$, then from eq. (7), $V_p$ is given by:

$$(16) \qquad\qquad V_p^2 = k/\varrho .$$

Jeffrey's original velocity distribution shows a discontinuous jump across the boundary between $F$ and $G$ and to accomodate this $k$ would have to increase by 32 per cent—excluding the highly improbable case that the density decreases with depth. Assuming Gutenberg's velocity depth curves the effective increase in $k$ would be 23 per cent. On the other hand, as BULLEN [16] first pointed out in 1946, if the inner core $G$ is solid and thus capable of transmitting $S$ waves, eq. (16) is replaced by eq. (7) in the region $G$ and the increase in $V_p$ can be accounted for without violating his $(k, p)$ hypothesis.

---

(*) In Sect. **4** and **5**, $k$ should strictly be $k_s$, the adiabatic incompressibility. The subscript $s$ has however been dropped.

BULLEN and HADDON [17] have since revised Bullen's original model B and constructed a series of models based on the $(k, p)$ hypothesis. The revision was undertaken to take into account the latest values of the moment of inertia of the Earth, seismic $P$ velocities in the core, and data from the free oscillations of the Earth.

From the earlier data on the moment of inertia of the Earth and the $P$ velocity distribution in the core, the density just below a depth of 80 km in Bullen's original model B had to be nearly 3.9 g(cm)$^{-3}$. In the original model A, on the other hand, the density was only 3.6 g(cm)$^{-3}$ at a depth of 400 km. An outstanding feature of the new B type models of Bullen and Haddon is that the $(k, p)$ hypothesis no longer requires the high densities of the original model B in the outermost several hundred kilometers of the Earth—in fact one of the chief results of these revised models is that the differences between models of A and B type have tended to disappear.

In a later paper, BULLEN and HADDON [18] constructed a series of Earth models in which the excess $\Delta k$ of the incompressibility $k$ at the top of the Earth's core over the value of $k$ at the bottom of the mantle was given different assigned values. Their calculations indicate that, unless the assumed seismic velocities and density gradients are more seriously in error than expected, $\Delta k/k$ does not exceed 2 per cent—in fact the most probable value of $\Delta k/k$ is insignificantly different from zero. On the other hand TAKEUCHI and KANAMORI [19] have calculated an equation of state for the materials of the Earth using shock wave data, and found that the variations in the values of the incompressibility of the metals studied is too large (even at 4 Mbar) to support Bullen's compressibility pressure hypothesis. They found it difficult, however, to avoid the conclusion that the inner core is solid. They also found that the density of the Eearth's core is about $(1 \div 1.5)$ g(cm)$^{-3}$ less than the density of iron at corresponding pressures and temperatures.

## 5. – Chemical inhomogeneity in the Earth.

For brevity the term chemical inhomogeneity is used to also include inhomogeneity arising from phase changes. Assuming hydrostatic pressure (eq. (2)) and an adiabatic temperature gradient, we have (by definition):

$$k = \varrho\varphi ,$$

so that

$$\frac{\mathrm{d}k}{\mathrm{d}p} = \varphi \frac{\mathrm{d}\varrho}{\mathrm{d}p} + \varrho \frac{\mathrm{d}\varphi}{\mathrm{d}p} = \varphi \frac{\mathrm{d}\varrho}{\mathrm{d}p} - \frac{1}{g} \frac{\mathrm{d}\varphi}{\mathrm{d}r} .$$

For a chemically homogeneous region, $d\varrho/dp = \varrho/k$ and

(17)
$$\frac{dk}{dp} = 1 - g^{-1}\frac{d\varphi}{dr}.$$

At any point of a region, chemically homogeneous or not, at which $d\varrho/dr$ can be assumed to exist, we can write

$$\frac{dk}{dp} = \frac{-\varphi}{g\varrho}\frac{d\varrho}{dr} - \frac{1}{g}\frac{d\varphi}{dr},$$

*i.e.*

(18)
$$\frac{d\varrho}{dr} = \frac{-\eta g\varrho}{\varphi},$$

where

(19)
$$\eta = \frac{dk}{dp} + g^{-1}\frac{d\varphi}{dr}.$$

When $\eta = 1$, eq. (18) reduces to the Adams-Williamson equation and eqs. (17) and (19) are identical. $\eta$ is thus a measure of the departure from chemical homogeneity—it is the ratio of the actual density gradient to the gradient that would obtain if the composition were uniform. An excess temperature gradient normally reduces while chemical inhomogeneity increases, the value of $\eta$.

From eq. (19) it can be seen that $\eta$ depends on $dk/dp$, $g$ and $d\varphi/dr$. On Bullen's compressibility-pressure hypothesis, $dk/dp$ is slowly varying and lies between about 3 and 6 throughout most of the Earth's deep interior. Uncertainties in estimates of $g$ are not large, while values of $d\varphi/dr$ are immediately derivable from the $P$ and $S$ velocity distributions. It follows from eq. (19) that $\eta$ can be estimated in most parts of the Earth's deep interior within limits that can be assigned. Thus it is possible to estimate the degree of departure from chemical homogeneity in any given region, and to assess density gradients where the Adams-Williamson equation cannot be used.

In the lower 200 km of the mantle (region $D''$), the seismic velocity distributions of both JEFFREYS and GUTENBERG indicate that $d\varphi/dr \simeq 0$, so that $\eta \simeq dk/dp$. Bullen's model A gives $dk/dp \simeq 3$ in $D''$ indicating that the lower $(100 \div 200)$ km of the mantle are inhomogeneous. The inhomogeneity is not too severe, however—with $\eta = 3$ it contributes only an extra 0.2 g(cm)$^{-3}$ density increase through $D''$.

In the transition region $F$ between the outer and inner core, the Jeffreys' velocity distribution (characterized by a large negative $P$ velocity gradient, *i.e.* $d\varphi/dr \gg 0$) leads to a value of $\eta$ of 38 entailing a density increase of the

order of 3 g (cm)⁻³ through $F$. On the other hand, the Gutenberg velocity
distribution gives large negative values of $d\varphi/dr$ so that $\eta$ is significantly less
than unity (actually negative) implying an unstable distribution of mass. It
would appear that seismic velocity gradients much in excess of those in
regions $D$ and $E$ cannot exist in the Earth's deep interior. An infinite gradient (*i.e.* a velocity discontinuity) on the other hand, is not impossible since
then the range of depth of any instability would be zero.

TABLE I. – *Assumed seismic data and layering in the core.*

| Region | Range of $r$ (km) | $V_p$ (km/s) |
|:---:|:---:|:---:|
| $E'$ | 3470÷1810 | JEFFREYS [21] |
| $E''$ | 1810÷1660 | 10.03 |
| $F$ | 1660÷1210 | 10.31 |
| $G$ | 1210÷0 | 11.23 |

In Bolt's [6, 20] revision of Jeffreys' distribution of the velocity of $P$ waves
in the deep interiore of the Earth, the core is divided into four distinct regions
$E'$, $E''$, $F$ and $G$. The velocity distribution down to the bottom of $E'$ is
the same as that of Jeffreys for corresponding depths inside $E$. There are
discontinuous jumps in $V_p$ at the $E''$-$F$ and $F$-$G$ boundaries and $V_p$ is constant in $E''$, $F$ and $G$ (see Table I and Fig. 5).

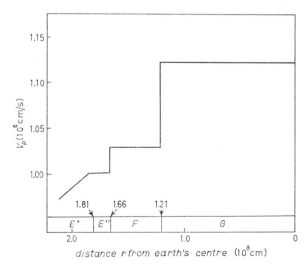

Fig. 5. – $P$ velocities and the regions $E', E'', F$ and $G$ of the lower core (after
K. E. BULLEN [23]).

BIRCH [22], using shock wave data at pressures of the order of $10^6$ atmospheres, inferred that the density $\varrho_0$ at the centre of the Earth does not exceed 13 g(cm)$^{-3}$. BULLEN [23] has investigated the consequences of this limiting value using Bolt's seismic $P$ velocity distribution in the core. Since $\varrho$ is not likely to decrease with depth in the core, Bullen's $(k, p)$ hypothesis implies, through eq. (7), that departures from smooth variations of $V_p$ with $r$ are accompanied by similar departures in the variation of $\mu$ rather than of $k$. BULLEN has shown that it is impossible for $\varrho_0$ to be as low as 13 unless there is substantial rigidity in both regions $F$ and $G$. In addition it is essential that $d\mu/dr > 0$ over a significant range of depth in the lower core, *i.e.*, the rigidity must *decrease* with increase of depth. BULLEN found that $\varrho_0$ could be as low as 12.6 g(cm)$^{-3}$ if suitable assumptions (compatible with the seismic data) are made on the variations of $k$ and $\mu$ in the lower core. The value of $\varrho_0$ could actually be reduced to 12.3 g(cm)$^{-3}$ if $E''$ were chemically homogeneous —which is, however, rather improbable. If $F$ and $G$ are both fluid (*i.e.* complete absence of rigidity), $\varrho_0$ must be at least 14.7 g(cm)$^{-3}$. This value is sufficiently in excess of 13 g(cm)$^{-3}$ to give additional support to the conclusion that the inner core is solid.

## 6. – The structure and composition of the Earth's mantle and core.

The first major discontinuity in the velocity depth curves (the Mohorovočić discontinuity or « Moho ») marks the boundary between the crust and mantle. Although the detailed structure of the crust is extremely complex, its mass is less than 1 per cent of that of the whole Earth and the details of its structure do not affect the interpretation of the physical properties of the deeper parts of the mantle and core. The depth to the Moho ranges from about 30 to 50 km below the continents but in general is not much more than 5 km beneath the ocean floors. At the Moho there is a discontinuous jump in the velocity of both $P$ and $S$ waves. It is not certain how « discontinuous » the velocity functions are but the change is thought to take place within a few km at most. There are two possibilities at the Moho—either it represents a chemical boundary separating solid phases having different chemical compositions or it is a phase boundary, separating two solid phases.

Seismic velocities in the crust are in the range of experimental velocities at kilobar pressures for common rocks such as granite and gabbro: below the Moho, seismic velocities are fairly consistent with those found experimentally for relatively rare dense rocks such as dunite, peridotite and eclogite. The generally accepted interpretation of the Moho is that it is a chemical discontinuity, separating crustal rocks characterized by a high feldspar content from underlying dunite or peridotite, essentially magnesium-iron olivines and pyroxenes. Two reactions have been suggested, however, as possible phase changes at the Moho—the gabbro-eclogite reaction and the olivine-serpentine reaction. These reactions—particularly the gabbro-eclogite reaction—have been studied extensively in recent years (see *e.g.* RINGWOOD and GREEN [24], GREEN and RINGWOOD [25]). The results indicate that the hypothesis that the Moho represents an isochemical transformation from gabbro to eclogite must be rejected—at least in normal continental regions—on a number of grounds. It is possible however that the transformation is important in some tectonically active areas *e.g.* continental margins, island arcs and mid-ocean ridges.

Studies of wave propagation in rocks show that the velocity of $P$ waves depends primarily upon density and mean atomic weight. However most common rocks have mean atomic weights close to 21 or 22 regardless of composition and rocks or minerals of very different composition may have the same densities and seismic velocities. Thus it is not easy to infer chemical composition from seismic data, and laboratory experiments at the conditions of pressure and temperature that exist deep within the Earth are highly desirable. Until very recently static high-pressure equipment using large presses could only reach a pressure of about 100 kbar which is equivalent to a depth of 300 km in the Earth. (In 1966 RINGWOOD and MAJOR [26] developed an apparatus capable of reaching pressures twice as great, but still only equal to those at a depth of about 600 km).

In the last few years, dynamic determinations of the compressibility of minerals and rocks have been made by a number of workers up to pressures in excess of those reached at the centre of the Earth. These high pressures are created for very short time intervals behind the front of a strong shock wave set up by an explosive charge and are an order of magnitude greater than pressures obtained by static methods. Numerical calculations have been carried out to determine equations of state for those materials likely to constitute the mantle and core—comparison of these equations of state with the experimental results indicates the type of materials which may exist in various regions of the Earth. The general results are in close agreement with the conclusions reached earlier by BIRCH and others, *viz.* that the mantle is made up of high-density oxides of silicon, iron, and magnesium, and that the core is mostly iron alloyed with some light elements. MCQUEEN *et al.* [27] found, for example, that a high-pressure modification of an olivine rock with a mean atomic weight of about 21.7 and an initial density of about 3.36 g(cm)$^{-3}$ satisfies the pressure-density and seismic requirements for the inner mantle.

6˙1. *The upper mantle (region B).* – The properties of the mantle are governed to an important degree by the response of the material to changes in pressure and temperature. Compositional variation affects the details of the overall picture but the main features are independent of composition, within the limits set by the cosmic adundances of the elements. The pressure distribution of the mantle is reasonably well established although temperatures are much less certain. At depths less than about 200 km the effect of temperature in changing the properties of the mantle is comparable to the effect of pressure because of high thermal gradients, and for certain properties the effects of temperature may be dominant. This is not the case at greater depths and throughout most of the mantle pressure is the dominant parameter.

There is now a growing body of evidence that there exists a low velocity (LV) layer at a depth of about 100 km in the upper mantle, thicker under ocean bottoms than under continents. Recent seismic results indicate that there are also significant regional differences. Beneath the oceans there is a definite LV zone for $S$ waves and possibly one for $P$ waves as well. Beneath Precambrian shields the LV zone for $S$ is less pronounced and the LV zone for $P$ seems to be absent. It is not known for certain what is the cause of the LV zone. It may represent a different mineral assemblage from the adjacent regions of the mantle or the material in this zone may be partially molten: it may also be caused by a thermal gradient so large that the effects of pressure are cancelled out. The LV zone is terminated fairly abruptly at a depth of about 150 km, indicating a sudden change in the physical state or composition of the material at that depth. The magnitude and abruptness of the velocity change favour a compositional change. Perhaps the lighter fraction of the mantle, which also has a lower melting point, has migrated upward, leaving behing a refractory residue which, not only has higher velocities, but is further from its melting point. The LV zone may represent a great reservoir of magma held in a solid matrix as water is held in a sponge. Since molten rock is enriched in radioactivity, a partially molten zone is self perpetuating. The conductivity of rock is so low that internally generated heat is effectively held in the Earth unless the molten rock is allowed to escape to the surface or to shallow depths.

6'2. *The transition region C*. – Surface wave studies (see *e.g.* ANDERSON [28, 29]) have shown that the abnormally high velocity gradients in the transition region are concentrated in two relatively narrow zones (50÷100) km thick instead of being spread out uniformly over some 600 km as was previously thought. Later studies, using travel times and the apparent velocities of body waves, have verified the presence of these two transition regions (see Fig. 6). In the most recent models of the upper mantle there is a very rapid increase in velocity between about 100 and 150 or 175 km and major discontinuities starting at (320÷365) km and (620÷640) km. The velocity gradients of the adjacent sections of mantle are appropriate for normal compression, including that section of the mantle between (440÷620) km which lies in the middle of what has been designated as the transition region. It has been shown that the locations of these transition regions, their general shape and their thicknesses are consistent with first, the transformation of magnesium-rich olivine to a spinel structure and then a further collapse of a material having approximately the properties of the component oxides.

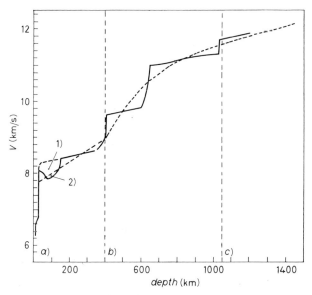

Fig. 6. – Distribution of seismic $P$ wave velocities in the outer 1200 km of the Earth according to H. JEFFREYS and C. B. ARCHAMBEAU, E. A. FLINN and D. G. LAMBERT. —— ARCHAMBEAU *et al.* ··· JEFFREY. *a*) Upper mantle. 1) Range of regional variations. 2) Low velocity zone. *b*) Transition zone. *c*) Lower mantle; Base of mantle, 2898 km, $V_p = 13.6$ km/s.

BIRCH [30, 1], using an equation of state based upon finite-strain theory, investigated the rate at which seismic velocities should increase with depth in a homogeneous medium with an arbitrary temperature gradient. He concluded that the properties of the upper mantle were consistent with the region being composed of familiar minerals such as olivines, pyroxenes and garnets. However the rate of increase of velocity with depth in region $C$ was too great for it to be homogeneous and BIRCH proposed that phase changes leading to close packing occurred in this region. Below 1000 km the elastic properties of the mantle were consistent with those of close-packed oxide phases which remainded unchanged to the core boundary. Although Birch's arguments, particularly those relating to phase transformation, met with some opposition, experimental investigations have since confirmed them.

The difficulty about testing Birch's hypothesis experimentally was that until 1963, available static high-pressure temperature apparatus was incapable of reproducing conditions in the Earth at depths greater than about

300 km. Thus prior to 1963 it was necessary to use *indirect* experimental methods based on thermodynamics, comparative crystal chemistry and, in particular, upon the study of germanate isotypes of silicates. This latter technique is due to the fortunate circumstance that germanates were often found to display the same kinds of phase transformations as the corresponding silicates, but at much lower pressures. (RINGWOOD [31] has given an excellent summary with a very complete bibliography of the results of this indirect phase of investigations which covered the period 1956-1963.) All of the germanate olivines and pyroxenes which were studied in the pressure range $(0 \div 90)$ kbar were found to be unstable at high pressures and transformed to dense phases suggesting that the corresponding silicate would transform similarly at higher pressures. During this period RINGWOOD found that the silicate olivines $Fe_2SiO_4$, $Ni_2SiO_4$ and $Co_2SiO_4$ could be transformed to spinel structures at $(20 \div 70)$ kbar and that the spinels were about 10 per cent denser than the corresponding olivines. A high-pressure form of silica $(SiO_2)$ having the rutile structure in which silicon was in octahedral co-ordination with a density of about 4.3 $g(cm)^{-3}$ was synthesized by STISHOV and POPOVA [32] at a pressure corresponding to a depth of about 100 km and has since been identified in the crushed zone in the Barringer Meteorite Crater, Arizona. This discovery suggested that the pyroxene family might transform to new structures such as ilmenite, characterized by octahedral co-ordination of $Si^4$.

In 1966 an apparatus capable of developing pressures above 200 kbar (corresponding to a depth of 600 km) simultaneously with high temperatures was developed by RINGWOOD and MAJOR [33, 26] in Australia. Other laboratories (in particular in Japan) have also now developed such techniques. Many new phase transformations of geophysical interest have been discovered both in silicates and in germanates—an excellent review has been given by RINGWOOD [34]. The rapid increase of seismic velocities around $(350 \div 450)$ km seems to be caused mainly by the transformation of pyroxenes into a new type of garnet structure and the transformation of olivines to the spinel (or related) structure. Our knowledge of the constitution of the mantle below 600 km rests mainly upon the interpretation of phase transformations in germanate analogue systems and on shock wave studies. These suggest that around $(600 \div 700)$ km, garnets and spinels transform to new phases possessing ilmenite, perovskite and strontium plumbate structures with densities and elastic properties resembling those of isochemical mixed oxides.

6˙3. *The lower mantle D.* – The net effect of the phase changes in the transition zone is to transform familiar silicate structures into close-packed structures similar to those of the dense oxides. An essential feature is the change in co-ordination of silicon from four-fold to six-fold. Once closest packing is attained, further transitions involving changes in crystal structure are not possible and the material remains homogeneous over a wide pressure range.

The constitution of the lower mantle has been discussed in detail by BIRCH [1, 8]. He concluded that the seismic velocity distribution is consistent with compression of a uniform layer, although the properties of this layer are not those of any known silicate. They resemble those possessed by relatively closed packed oxides such as corundum, periclase, rutile and spinel. There is no evidence suggesting further phase changes or significant changes in chemical composition. Since velocity and density data provide only two useful properties of the material in the lower mantle, only three compositional variables at most can be fixed. Consideration of meteorite and solar abundances indicates that the dominant oxides in the earth are $SiO_2$, MgO and FeO, and thus these are the ones whose ratios shoud be adjusted. The abundance figures allow small amounts of other oxides such as $Al_2O_3$, CaO and $Na_2O$, but they will not have a large effect. Taking the phases in the lower mantle to be an oxide solid solution with the periclase structure and a metasilicate solid solution with the ilmenite structure, CLARK and RINGWOOD [35] have shown that a molecular ratio of $MgO/SiO_2$ of 1.5 and a molecular ratio of $FeO/(FeO+MgO)$ of about 0.1 provide a good fit to the density and velocity data. These ratios are also very plausible on geochemical grounds.

CHINNERY and TOKSÖZ [36] and TOKSÖZ et al. [37] have determined $P$ wave velocities in the lower mantle from $dT/d\Delta$ measurements using the large aperture seismic array in Montana and travel times from the LONGSHOT nuclear explosion. Previous models for the $P$ wave velocity profile were based primarily on observations of travel times and amplitudes from earthquakes—with the inherent lack of accurate origin times and epicentres and the variability of measured amplitudes. They found that the velocity structure showed anomalous gradients or «discontinuities» at depths of about 1200 and 1900 km. It is not possible to say whether these discontinuities are global since the data used came from one small part of the Earth, and there is evidence that lateral inhomogeneities persist to considerable depth. Discontinuities in the upper mantle at depths of about 300 and 700 km were detected earlier and explained as possible phase changes (see Subsect. 6'2). The discovery of discontinuities at depths of about 1200 and 1900 km indicates regions where there are either phase or composition changes or both, and, contrary to earlier beliefs, point to departures from homogeneity in the lower mantle.

Using the extended array at the Tonto Forest Seismological Observatory in Arizona to measure $dT/d\Delta$ of direct $P$ waves in the distance range $(30\div100)°$, JOHNSON [38] estimated $P$ velocities in the lower mantle. He found no large discrepancies with the traditional model of Gutenberg and Jeffreys, but increased velocity gradients near depths of 830, 1000, 1230, 1540, 1910 and 2370 km were indicated. It must be stressed however, that these anomalies in the lower mantle are spread over depth intervals of at least 50 km, and are an order of magnitude smaller than those which have been found for the upper mantle.

6˙4. *The core.* – There has been much speculation on the origin and evolution of the Earth's core, which is probably bound up with the origin of the Earth itself. It is a common assumption in most theories of the origin of the Earth that the proto-Earth was homogeneous and that the present differentiation into a core, mantle and crust occurred late. There are, however, as RINGWOOD [39] has pointed out, a number of difficulties with such a model, and there have been some qualitative considerations in recent years of a non-homogeneous accretion of the Earth and planets.

OROWAN [40] pointed out that iron is plastic-ductile, even at low temperatures, provided that it does not contain far more carbon than is found in meteorites. If it is assumed that the planets have agglomerated from solid particles condensed from a gaseous atmosphere around the sun, metallic particles would be expected to stick together when they collide because they can absorb kinetic energy by plastic deformation. They can therefore unite by « cold welding » or by hot welding. Silicates, on the other hand, are brittle and break up in a collision except within a narrow temperature range near their melting point. The agglomeration of the planets may thus start with metallic particles. When the body, built up chiefly from heavy metal particles, is sufficiently large it can easily collect nonmetallic particles by embedding them in ductile metal, and later by its gravitational attraction on the fragments resulting from collisions. OROWAN thus suggests that planets may arise cold in this way with a metal core already partially differentiated, and that subsequent melting can produce a sharp boundary between core and mantle.

TUREKIAN and CLARK [41] have also considered a model of the planets stratified initially due to the inhomogeneous accumulation of the elements. As the primitive solar nebula cooled, elements and compounds would condense in the order of increasing vapour pressure. Assuming a pressure between $10^{-3}$ and 1 atmosphere, LARIMER [42] calculated that the order of condensation would be iron and nickel, magnesium and iron silicates, alkali silicates, metals such as Ag, Ga, Cu, iron sulphide and finally metals such as Hg, Tl, Pb, In and Bi. This order of condensation is grossly that inferred in the Earth. Such stratification is usually attributed to the settling of the densest material to the center. High density however is associated with low volatility enabling planets to accrete in a manner that is automatically gravitationally stable. TUREKIAN and CLARK thus suggest that the Earth's core formed by accumulation of the condensed iron-nickel in the vicinity of its orbit—and then became the nucleus upon which the silicate mantle was deposited. The last accumulates would be FeS, $Fe_3O_4$, the volatile trace elements, organic compounds, hydrated silicates and rare gases.

It has been shown (LATIMER [43]; UREY [44]) that all of the common metals in a gas and dust cloud of solar composition would occur in the form of *oxides* at temperatures below 300 °K under equilibrium conditions. This conclusion is of the utmost importance in the case of iron. Since *metallic* iron is an important constituent in meteorites and of the Earth it appears that accretion of the primitive dust in the large bodies was accompanied or preceded by chemical reduction of iron and nickel oxides to the metallic state. RINGWOOD [45] proposed that the Earth formed directly by accretion from the primitive oxidized dust in the solar nebula, and that reduction to metal, loss of volatiles, melting and differentiation occurred simultaneously and as a direct result of the primary accretion process. During the later stages of accretion when the melting point of the surface was exceeded, metallic iron segregated into masses large enough to flow directly into the core. Gravitational energy liberated during the formation of the core would also contribute heat sufficient to result in complete melting throughout. On this model segregation of the core occurred as a continuous process during the primary accretion. A catastrophic version of core formation is also a possibility. According to Ringwood's model the Earth developed in a state which was grossly out of chemical equilibrium. The deep interior was initially highly oxidized and rich in volatiles whereas the outer regions were progressively more reduced and poor in volatile components. After melting near the surface, the metal phase consisting of an iron-nickel-silicon alloy collected into bodies which were large enough to sink into the core.

TOZER [46] has reconsidered the question of the kinetics of core formation and calculated that the simple theory of falling iron masses in silicate material, as considered by ELSASSER [47] for example, is untenable. He suggests as an alternative a mechanism based on the flow of iron along channels in the silicate phase—the acceptability of this theory however depends quite critically on whether iron is able to flow over distances of the order of a km under such conditions. TOZER concludes, however, that in any case core formation is proceeding today much more slowly than in the past and that it was virtually complete very early in the Earth's history.

In 1949 RAMSEY [48] suggested that the discontinuity at the core-mantle boundary might be due to a high-pressure phase change. He thus visualized a chemically homogeneous composition for the earth (below the crustal layers) which he identified as olivine; at the core-mantle boundary there would be a transition from a molecular to a metallic phase. Originally RAMSEY put forward this theory to account for the differences in the observed densities of the terrestrial planets which he assumed to have a common primitive composition. More recent astronomical data have revised the older figures and make Ramsey's hypothesis improbable. Moreover shock wave experiments lend no support to the hypothesis but suggest that the inner core is of nickel-iron composition, although the outer core may possibly consist of a modification of ultra-basic rock. Finally a phase change in a multicomponent system almost certainly would be spread over a range of pressure, whereas the core-mantle boundary presents a very sharp discontinuity.

Studies of the physical properties of the Earth's core by BIRCH [1] have indicated that it is about 10 per cent less dense than nickel-iron and that its seismic velocity is substantially greater than that of nickel-iron under comparable conditions of temperature and pressure. These conclusions have been confirmed by recent investigations on the densities and seismic velocities of metals at pressures equal to those in the core using shock wave techniques (see Fig. 7 and BIRCH [22]). Unfortunately shock wave data supply only one

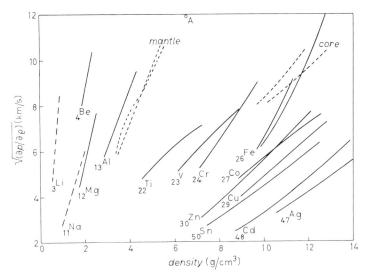

Fig. 7. – Hydrodynamical velocity, $(\partial p/\partial \varrho)^{\frac{1}{2}}$, vs. density. The solid curves are shock wave data for metals; the dashed lines are for static compressions. The dashed curves for mantle and core are obtained from seismic velocities combined with representative density distributions. The circle labeled « A » is for dunite at 2.4 Mbar (after F. BIRCH [22]).

elastic constant—the compressibility—and the effects of temperature can only be treated approximately. Silicates are structurally complicated and the interatomic forces between constituent ions are not as clear cut as in simple ionic crystals, so that a complete lattice dynamical or quantum mechanical calculation for any common rock forming mineral has not yet been carried out. Even pressures in the core are too low for the applicability of simplified statistical treatments such as the Thomas-Fermi-Dirac equation of state.

Figure 7 is a graph of the hydrodynamical sound velocity (which is related to the seismic velocities) against density. The solid curves are experimental and the broken curves show the same quantities for the mantle and core. No adjustment of the data will allow a core of light metals or their oxygen compounds, nor can the mantle be made of heavy metals. Transformation of light compounds to a metallic state may take place in the Earth but the density of the core requires a metal of the transition group and only iron is sufficiently abundant. The properties of iron are close to those required and can be adjusted with small amounts of light alloying elements. It appears therefore that the Earth's core contains a substantial amount of an element with a low density which can also increase the elastic ratio and seismic velocity of iron. Limitations upon possible choices are that the element must be reasonably abundant, miscible with liquid iron and possess chemical properties which will allow it to enter the core. RINGWOOD [49] has shown that silicon is the most likely extra component of the Earth's core. At the mantle core boundary magnesium oxide could also be soluble to an extent of about 10 per cent in liquid iron. Such solubility considerations put stringent constraints on the chemical constitution of the Earth—ALDER [50] has shown that neither $SiO_2$ nor $Mg_2SiO_4$ are sufficiently soluble to lower the density of the liquid core significantly below that of pure iron.

Many calculations have been carried out in an attempt to estimate the loss of gravitational energy and the consequent rise of temperature associated with core formation. BIRCH [51] has simplified the problem by assuming that the undifferentiated Earth consists of a homogeneous mixture of the materials of the present core and mantle. Only simple unmixing was considered, the total mass remaining unchanged, so that his calculations do not include an allowance for chemical reactions or loss of volatile components. Without allowance for thermal expansion, BIRCH found that the mean energy available for heating is 600 cal/g; with an approximate allowance for thermal expansion this is reduced to 400 cal/g which is equivalent to a mean rise of temperature of about 1600 °C. This suggests that core formation within an originally homogeneous Earth would be the most important happening in its thermal history after accretion—it may in fact represent the 4.5 billion year age of the Earth.

Although we have no direct knowledge of the time of origin of the Earth's molten metallic core, nevertheless, if it is assumed that the Earth's magnetic field results from motions in a predominantly fluid iron core, a time constraint can be placed on core formation. It must have antedated the oldest known rock possessing remanent magnetism. The age of some rocks in Africa possessing remanent magnetism is (2.5÷2.7) billion years. Thus if the Earth is 4.5 billion years old then core formation must have occurred within the first 2.0 billion years of the Earth's history. HANKS and ANDERSON [52] have carried out a series of calculations on the early thermal history of the Earth taking into account this boundary condition. The results of their calculations are that large scale differentiation within the Earth leading to core formation must almost certainly have taken place during an accretion period lasting 500,000 years or less. In other words, the origin of the Earth and its core would have taken place practically simultaneously.

Recently HIDE [53] has considered the topography of the core mantle interface and suggsted that « bumps » on the boundary (vertical dimensions of the order of a few kilometers), too small to be observed by modern seismological methods, might nevertheless produce pronounced hydrodynamical effects in the core. In a later paper HIDE and HORAI [54] constructed hypothetical contour maps of the core-mantle interface based on the assumption that the low degree harmonics of the Earth's gravitational field are associated with undulations at the core-mantle boundary. DOELL and COX [55] found from palaeomagnetic evidence that the secular variation over the Pacific hemisphere has been systematically weaker than elsewhere on the Earth's surface for the past million years and HIDE [56] has attempted to explain this also by the interaction of fluid motions in the core with topographical features of the core-mantle interface.

HIDE and MALIN [57] and KHAN and WOOLLARD [58] have sought correlations between global features of the Earth's gravitational and magnetic fields. In particular HIDE and MALIN claim that there is a significant correlation between the Earth's gravitational field and the nondipole part of the geomagnetic field if the latter is displaced about 160° in longitude. The statistics however do not appear too convincing. Nevertheless the possibility of the existence of « bumps » on the core-mantle boundary is important—particularly the question of the time scale on which such features can persist and whether they can grow and decay.

## APPENDIX A

In a chemically homogeneous layer,

$$\mathrm{d}\varrho = \left(\frac{\partial \varrho}{\partial p}\right)_T \mathrm{d}p + \left(\frac{\partial \varrho}{\partial T}\right)_p \mathrm{d}T = \frac{\varrho}{k_T}\, \mathrm{d}p - \varrho\alpha\, \mathrm{d}T\,,$$

where $k_T$ is the isothermal incompressibility and $\alpha$ the (volume) coefficient of thermal expansion. Using eq. (2) this becomes:

$$\frac{\mathrm{d}\varrho}{\mathrm{d}r} = \frac{-g\varrho^2}{k_T} - \varrho\alpha\frac{\mathrm{d}T}{\mathrm{d}r}\,.$$

The thermal gradient can be split into an adiabatic gradient and a nonadiabatic component $\tau$ (considered positive when the increase of temperature with depth exceeds the adiabatic).

Since $(\partial T/\partial p)_s = T\alpha/\varrho c_p$ where $c_p$ is the specific heat at constant pressure, it follows that we can write:

$$\frac{\mathrm{d}T}{\mathrm{d}r} = \frac{-Tg\alpha}{c_p} - \tau \; .$$

Hence

$$\frac{\mathrm{d}\varrho}{\mathrm{d}r} = \frac{-g\varrho^2}{k_T} + \frac{T\varrho\alpha^2 g}{c_p} + \alpha\varrho\tau \; .$$

Since

$$\frac{k_T}{k_s} = 1 - \frac{T\alpha^2 k_T}{\varrho c_p}$$

this simplifies to

$$\frac{\mathrm{d}\varrho}{\mathrm{d}r} = \frac{-g\varrho^2}{k_s} + \alpha\varrho\tau = \frac{-g\varrho}{\varphi} + \alpha\varrho\tau \; .$$

## REFERENCES

[1] F. BIRCH: *Journ. Geophys. Res.*, **57**, 227 (1952).
[2] K. E. BULLEN: *An Introduction to the Theory of Seismology*, 3rd edition (London, 1963).
[3] M. LANDISMAN, Y. SATO and J. NAEF: *Geophys. Journ.*, **9**, 439 (1965).
[4] K. E. BULLEN: *Nature*, **196**, 973 (1962).
[5] K. E. BULLEN: *Geophys. Journ.*, **7**, 584 (1963).
[6] B. A. BOLT: *Nature*, **196**, 122 (1962).
[7] C. L. PEKERIS: *Geophys. Journ.*, **11**, 85 (1966).
[8] F. BIRCH: *Journ. Geophys. Res.*, **69**, 4377 (1964).
[9] S. P. CLARK jr. and A. E. RINGWOOD: *Rev. Geophys.*, **2**, 35 (1964).
[10] R. A. W. HADDON and K. E. BULLEN: *Phys. Earth Planet. Int.*, **2**, 35 (1969).
[11] F. PRESS: *Science*, **160**, 1218 (1968).
[12] F. PRESS: *Journ. Geophys. Res.*, **73**, 5223 (1968).
[13] F. PRESS: *Phys. Earth Planet. Int.*, **3**, 3 (1970).
[14] K. E. BULLEN: *Mont. Not. Roy. Astr. Soc. Geophys. Suppl.*, **5**, 355 (1949).
[15] K. E. BULLEN: *Mont. Not. Roy. Astr. Soc. Geophys. Suppl.*, **6**, 50 (1950).
[16] K. E. BULLEN: *Nature*, **157**, 405 (1946).
[17] K. E. BULLEN and R. A. W. HADDON: *Phys. Earth Planet. Int.*, **1**, 1 (1967).
[18] K. E. BULLEN and R. A. W. HADDON: *Geophys. Journ.*, **17**, 179 (1969).
[19] H. TAKEUCHI and H. KANAMORI: *Journ. Geophys. Res.*, **71**, 3985 (1966).
[20] B. A. BOLT: *Bull. Seism. Soc. Amer.*, **54**, 191 (1964).
[21] H. JEFFREYS: *Mont. Not. Roy. Astr. Soc. Geophys. Suppl.*, **4**, 594 (1939).

[22]  F. BIRCH: *Solids Under Pressure*, edited by W. PAUL and D. M. WARSCHAUER (New York, 1963), p. 137.

[23]  K. E. BULLEN: *Geophys. Journ.*, **9**, 233 (1965).

[24]  A. E. RINGWOOD and D. H. GREEN: *Tectonophysics*, **3**, 383 (1966).

[25]  D. H. GREEN and A. E. RINGWOOD: *Geochim. Cosmochim. Acta*, **31**, 767 (1967).

[26]  A. E. RINGWOOD and A. MAJOR: *Phys. Earth Planet. Int.*, **1**, 164 (1968).

[27]  R. G. McQUEEN, S. P. MARSH and J. N. FRITZ: *Journ. Geophys. Res.*, **72**, 4999 (1967).

[28]  D. L. ANDERSON: *Physics and Chemistry of the Earth*, Vol. 6 (London, 1966), p. 1.

[29]  D. L. ANDERSON: *The Earth's Mantle*, edited by T. F. GASKELL (New York, 1967), p. 355.

[30]  F. BIRCH: *Bull. Seism. Soc. Amer.*, **29**, 463 (1939).

[31]  A. E. RINGWOOD: *Earth Planet. Sci. Lett.*, **5**, 401 (1969).

[32]  S. M. STISHOV and S. V. POPOVA: *Geokhimiya*, **10**, 837 (1961).

[33]  A. E. RINGWOOD and A. MAJOR: *Earth Planet. Sci. Lett.*, **1**, 241 (1966).

[34]  A. E. RINGWOOD: *Phys. Earth Planet. Int.*, **3**, 109 (1970).

[35]  S. P. CLARK jr. and A. E. RINGWOOD: *The Earth's Mantle*, edited by T. GASKELL (New York, 1967), p. 111.

[36]  M. A. CHINNERY and M. N. TOKSÖZ: *Bull. Seism. Soc. Amer.*, **57**, 199 (1967).

[37]  M. N. TOKSÖZ, M. A. CHINNERY and D. L. ANDERSON: *Geophys. Journ.*, **13**, 31 (1967).

[38]  L. R. JOHNSON: *Bull. Seism. Soc. Amer.*, **59**, 973 (1969).

[39]  A. E. RINGWOOD: *Geochim. Cosmochim. Acta*, **30**, 41 (1966).

[40]  E. ORWAN: *Nature*, **222**, 867 (1969).

[41]  K. K. TUREKIAN and S. P. CLARK jr.: *Earth Planet. Sci. Lett.*, **6**, 346 (1969).

[42]  J. W. LARIMER: *Geochim. Acta*, **31**, 1215 (1967).

[43]  W. M. LATIMER: *Sience*, **112**, 101 (1950).

[44]  H. C. UREY: *The Planets* (Yale Univ. Press, 1952).

[45]  A. E. RINGWOOD: *Geochim. Cosmochim. Acta*, **20**, 241 (1960).

[46]  D. C. TOZER: *Geophys. Journ.*, **9**, 95 (1965).

[47]  W. M. ELSASSER: *Earth Science and Meteoritics*, edited by J. GEISS and E. D. GOLDBERG (Amsterdam, 1963), p. 1.

[48]  W. H. RAMSEY: *Mon. Not. Roy. Astr. Soc. Geophys. Suppl.*, **5**, 409 (1949).

[49]  A. E. RINGWOOD: *Advances in Earth Sciences*, edited by P. M. HURLEY (Cambridge, Mass., 1966), p. 287.

[50]  B. J. ALDER: *Journ. Geophys. Res.*, **71**, 4973 (1966).

[51]  F. BIRCH: *Journ. Geophys. Res.*, **70**, 6217 (1965).

[52]  T. C. HANKS and D. L. ANDERSON: *Phys. Earth Planet. Int.*, **2**, 19 (1969).

[53]  R. HIDE: *Science*, **157**, 55 (1967).

[54]  R. HIDE and K. HORAI: *Phys. Earth Planet. Int.*, **1**, 305 (1968).

[55]  R. R. DOELL and A. COX: *Journ. Geophys. Res.*, **70**, 3377 (1965).

[56]  R. HIDE: *Phil. Trans. Roy. Soc., Lond. Ser.*, A **259**, 615 (1966).

[57]  R. HIDE and S. R. C. MALIN: *Nature*, **225**, 605 (1970).

[58]  M. A. KHAN and G. P. WOLLARD: *Nature*, **226**, 340 (1970).

# Equations of State and the Interior of the Earth.

L. THOMSEN

*Laboratoire des Hautes Pressions - Bellevue*

## 1. – Introduction.

An equation of state, for our purposes, is a relation between the pressure, the specific volume, and the temperature of a material. We have thus eliminated at the outset any consideration of (finite) nonhydrostatic stresses. Further, we have ignored the details of any nonisotropic strain response to the assumed hydrostatic pressure. The first simplification is justified on two grounds: 1) « large » nonhydrostatic stresses lead, in most materials (excepting rubber, etc.) to irreversible (*e.g.* plastic) deformations, hence they cannot be included strictly in a discussion of the equation of state, which must not involve the history of the material; 2) this property of materials may be rephrased as the statement that the materials of the Earth will not support nonhydrostatic stresses over geologic times, *i.e.*, have long since yielded to them; hence the only « large » stresses of geophysical interest are pure pressures. Here « large » means greater than the shear strength of the medium, hence this statement must be modified in the outer layers of planets (*e.g.* some tens of kilometers for the Earth) where relatively low temperatures allow the shear strengths and stresses to be appreciable in some instances. The second simplification is justified by the seismological verification that the material of the Earth's mantle, while certainly anisotropic on a microscopic scale, exists in the mantle to a good approximation as a randomly oriented mass of small crystals, which when averaged over a typical seismic wave-length, is effectively isotropic. The relation between the elastic properties of the single crystals and those of the polycrystalline mass is not direct [2], but we shall avoid the issue here, restricting the discussion to formally isotropic media, and to cubic crystals.

Furthermore, we shall limit ourselves to a restricted portion of the *P-T* plane, one corresponding to the conditions encountered by the terrestrial planets. Figure 1 illustrates this regime schematically, and indicates the complexity of a more general discussion. It develops that this regime is one of the more difficult to study. On the one hand, such *high* pressures and temperatures are difficult to generate experimentally. The problems associated with even our

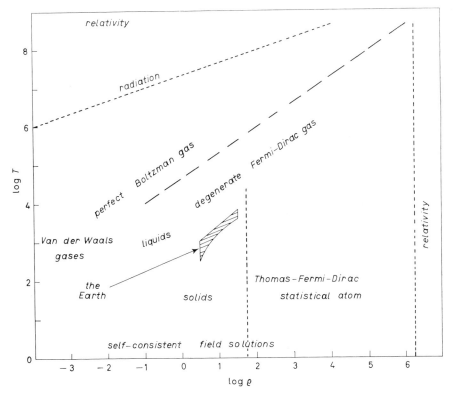

Fig. 1. – A semi-quantitative diagram of the $\varrho - T$ fields where various idealizations of the state of matter are valid. The definition of the fields in the lower left-hand portion is intentionally vague, due to the complexity of the physics in this region.

most sophisticated efforts, the strong shock waves, are discussed by KNOPOFF and by AHRENS elsewhere in these lectures. On the other hand, such *low P* and *T* involve the details of crystal structure, quantum levels etc. in important ways, rendering invalid the theoretical models mentioned in the very high *P* and *T* regions of Fig. 1. Again KNOPOFF has indicated how this situation limits the geophysical utility of the TFD and Wigner-Seitz theories. One helpful aspect is that the temperatures in the regime of interest are generally above the characteristic Debye temperatures of the materials involved, so that their vibrational properties may be treated classically. This statement will be given precision below.

## 2. – Thermodynamics.

Most equations of state are simply statements of the conservation of energy. The first law of thermodynamics, in its most familiar form is

$$(1) \qquad\qquad \mathrm{d}U(S,\, V) = T\,\mathrm{d}S - P\,\mathrm{d}V\,,$$

where $U$ is the *total* energy of a system. It is convenient, from a *theoretical* point of view, to choose the independent variables as $T$ and $V$. This is because, respectively, it is the *temperature* (rather than the entropy) which defines the basic phenomen of thermal equilibrium, and because the basic description of a physical system generally starts in terms of *spatial* variables (rather than the stresses, which are generally boundary conditions). Hence it is convenient to write the first law in terms of the Helmholtz *free* energy, $F = U - TS$:

$$(2) \qquad\qquad \mathrm{d}F(T,\, V) = -S\,\mathrm{d}T - P\,\mathrm{d}V\,.$$

In solid-state *experiments*, it is usually the *pressure* rather than the volume, which is arbitrarily adjustable; also acoustic propagation is *adiabatic* rather than isothermal. While this leads to formulae which may appear awkward, no confusion need result.

The partial differential forms of (2) are

$$(3) \qquad\qquad S = -\left(\frac{\partial F}{\partial T}\right)_V$$

and

$$(4) \qquad\qquad P = -\left(\frac{\partial F}{\partial V}\right)_T.$$

The first is called the *caloric* equation of state; the second is called the *thermal* equation of state, and will receive most of the following discussion. The depth of this bifurcation will not appear evident until somewhat later.

$2\,{}^{\cdot}1$. « *Finite-strain theory* » [3]; *the Lagrangian formulation*. – The study of the equation of state reduces to a determination of the free energy $F(V,\, T)$. The exact solution of this problem is prohibited by the intricacies of Schrö-dinger's equation, and by the magnitude of Avogadro's number. However, some imaginative efforts have been made to bypass these fundamental dif-ficulties, and to determine the free energy from purely geometrical arguments. The general idea here is deceptively simple: $F$ may obviously be expanded around its zero-pressure value in a Taylor series in the elastic « strain ».

$$(5) \qquad\qquad F = \sum_{i=0}^{\infty} A_i \, (\text{« strain »})^i \,.$$

The « strain » parameter contains all the information concerning the change in volume, the coefficients $A_i$ contain all the physics, all the quantal and many-body difficulties mentioned above, and cannot be evaluated from first principles. But they *can be* simply regarded as *i*-order derivatives of $F$ (evaluated at zero « strain »), and determined by *experiments* giving the initial pressure (zero), the initial incompressibility ($K_0$), etc. If the series converges quickly, one may hope to produce in this manner an accurate equation of state.

This idea proceeds well enough initially. We first define finite « strain », *i.e. a tensor transformation which*:

1) *connects the two vector fields describing the initial and deformed configurations, and*

2) *reduces, in the infinitesimal limit, to the familiar infinitesimal strain tensor of acoustics.*

It is seen that this general definition allows considerable arbitrariness to persist; we refer to « strain » with quotation marks so as to remind ourselves of this ambiguity. Let us proceed with caution. We let $R_0$ be the vector to a general point in the natural configuration, and $R$ be the vector to the same point in the deformed configuration. Then we may *define* a finite « strain » tensor $e$ by

$$(6) \qquad\qquad R_i \equiv (\mathbf{1} + e)_{im} R_{0m} \,.$$

Here the unit tensor $(\mathbf{1})_{im} = \delta_{im}$ (the Kronecker delta) is included explicitly in the trasformation matrix so that $e$ will satisfy the second part of the definition above. If we write the *displacement* $\boldsymbol{u} = \boldsymbol{R} - \boldsymbol{R}_0$, then (6) becomes

$$u_i \equiv e_{im} R_{0m}$$

and it is clear that $e_{im} = \partial u_i / \partial R_{0m}$ as in infinitesimal elasticity. Thus (6) represents a simple-minded extension of infinitesimal ideas into the finite regime. It is called a « Lagrangian » or material definition of strain; because it transforms the undeformed state $[\boldsymbol{R}_0]$ into the deformed state $[\boldsymbol{R}]$.

The tensor $e$ is not suitable as an expansion parameter in (5), however, for the following reason. We assume that the free energy of a body is invariant towards pure rotations of the body, hence the « strain » tensor used to describe its deformation must be similarly invariant. This is true even in the present case of hydrostatic pressure (hence no rotation); it is a property of the body itself, and not of the process. To find a rotationally invariant equivalent of (6), we consider the *distance*, $R^2$:

$$R^2 = R_i R_i = (1 + e)_{im} R_{0m} (1 + e)_{in} R_{0n} =$$
$$= R_{0m} (1 + e^\dagger)_{mi} (1 + e)_{in} R_{0n} = R_{0m} (1 + 2\eta)_{mn} R_{0n} .$$

Here the dagger represents the transposed tensor, $e^\dagger_{mi} = e_{mi}$. We have introduced a new tensor $\eta$ defined by

(7)                    $$(1 + 2\eta) \equiv (1 + e^\dagger)(1 + e) ,$$

or

(8)                    $$\eta = \tfrac{1}{2}(e + e^\dagger + e^\dagger e) ,$$

which is symmetric, hence rotationally invariant. It is clear that the quadratic term in (8) appears because of the rotational invariance, and not as the start of an infinite series.

With the assumption of homogeneous deformation, the relation of $\eta$ to the change in volume is very simple. The Jacobian of the simple transformation (6) is

$$J = \frac{V}{V_0} = |\det (1 + e)| .$$

Through simple identities, we have also

$$\frac{V}{V_0} = \left|\det (1 + e^\dagger)\right| = \sqrt{\left|\det (1 + e^\dagger)(1 + e)\right|}$$

and

(9)                    $$\left(\frac{V}{V_0}\right)^2 = |\det (1 + 2\eta)| .$$

In our case of hydrostatic pressure on a cubic or isotropic material, $(1 = 2\eta)_{mn} = (1 + 2\eta)\delta_{mn}$, and from (9),

(10)                   $$\eta = \frac{1}{2}\left[\left(\frac{V}{V_0}\right)^{\frac{2}{3}} - 1\right] .$$

This scalar varies from $0$ to $-\tfrac{1}{2}$ as the pressure varies from 0 to $\infty$. It may now be substitued into (5):

$$F = \sum_0^\infty A_i \eta^i .$$

With eq. (4) for the pressure, $K = -V(\partial P/\partial V)_T$ for the isothermal incompressibility, $K' = (\partial K/\partial P)_T$, for the first pressure derivative, etc., the first few coefficients may be evaluated:

(11a)
$$A_0 = A_1 = 0 \,,$$

(11b)
$$A_2 = \frac{9}{2} K_0 V_0 \,,$$

(11c)
$$A_3 = -\frac{9}{2} K_0 V_0 K_0' \,,$$

(11d)
$$A_4 = \frac{9}{2} K_0 V_0 \left[ K_0 K_0'' + K_0'(K_0' + 1) - \frac{1}{9} \right] \,.$$

Here the subzero indicates evaluation at $P = 0$, $T = T^0 =$ room temperature. The equation of state appears as:

(12)
$$P = -3K_0 \left( \frac{V}{V_0} \right)^{-\frac{5}{3}} \left[ \eta - \frac{3}{2} K_0' \eta^2 + \ldots \right] \,.$$

It is impossible to evaluate the higher-order terms experimentally; already $A_4$ presents difficulties, and must usually be evaluated indirectly (see Subsect. 2˙4). So what we have is a small sampling of an infinite set of terms, and it is difficult to say anything definite about the effect of neglecting those of higher order. Their size must be estimated on the basis of theories on the atomic forces, yet it is precisely such theories which we had hoped to avoid by this geometrical continuum mechanics approach. Experimentally, the $|A_i|$ appear to increase with $i$ as far as we can see, so we may presume the radius of rapid convergence of the series (12) to be quite restricted. (However such a presumption has no *physical* justification; the only *physical* input is the conservation of energy and the rotational invariance of $F$.)

2˙2. *The Eulerian Formulation.* – In an attempt to use the previously mentioned ambiguity in the definition of « strain » to resolve this impasse, MURNAGHAN [3a] adopted different definitions. He defined a « strain » tensor $f$ by

(13)
$$R_{0i} = (1 - f)_{im} R_m$$

(compare with (6)). This definition is essentially a transformation of the deformed configuration $[R]$ back into the initial one $[R_0]$; although it is perhaps less straightforward than (6), it is formally correct. It is an « Eulerian », or spatial, definition of strain, in contrast to the Lagrangian definition (6).

The associated symmetric « strain » tensor is

(14)
$$\boldsymbol{\epsilon} = \tfrac{1}{2}(\boldsymbol{f} + \boldsymbol{f}^\dagger - \boldsymbol{f}^\dagger \boldsymbol{f})$$

(compare with (8)) and the associated scalar $\varepsilon$ is related to the volume change by

(15)
$$\varepsilon = \frac{-1}{2}\left[\left(\frac{V}{V_0}\right)^{-\frac{2}{3}} - 1\right]$$

(compare with (9)).

The fact that $(V/V_0)^{-\frac{2}{3}}$ appears in (15), as opposed to $(V/V_0)^{+\frac{2}{3}}$ in (9) is a reflection of the difference in viewpoint adopted in (13); it means that as $P \to \infty$, $\varepsilon \to -\infty$ instead of being bound, as $\eta$. Although, we are not concerned with this *limiting* behaviour, we shall see that the two viewpoints *do* lead to significantly different results. Making an expansion of the free energy as before

$$F = \sum_0^\infty B_j \varepsilon^j \,,$$

the first few coefficients are:

(16a)
$$B_0 = B_1 = 0 \,,$$

(16b)
$$B_2 = A_2 = \frac{9}{2} K_0 V_0 \,,$$

(16c)
$$B_3 = -\frac{9}{2} K_0 V_0 (K_0' - 4) \,,$$

(16d)
$$B_4 = \frac{9}{2} K_0 V_0 \left[ K_0 K_0'' + K_0'(K_0' - 7) + \frac{143}{9} \right]$$

and the equation of state appears as

(17)
$$P = -3K_0 \left(\frac{V}{V_0}\right)^{-\frac{5}{3}} \left[ \varepsilon - \frac{3}{2}(K_0' - 4)\varepsilon^2 + \dots \right].$$

This equation has achieved prominence in the geophysical literature as the « Birch-Murnaghan equation of state » [5]. The *infinite* expansion (17) is presumably identical to the *infinite* expansion (12), but an easy calculation reveals that the *truncated* series differ. Figure 2 shows the two curves for CsI (for which $A_4$, $B_4$ may be directly evaluated); the difference becomes considerable. One sees that the higher-order terms for one or both of these formulations must be important. The experimental evidence, in this case, tends to confirm [6] the predictions of the fourth-order Lagrangian formulation, eq. (12), though no general conclusions can be drawn yet from this result. See, however, Subsect. **3**'4.

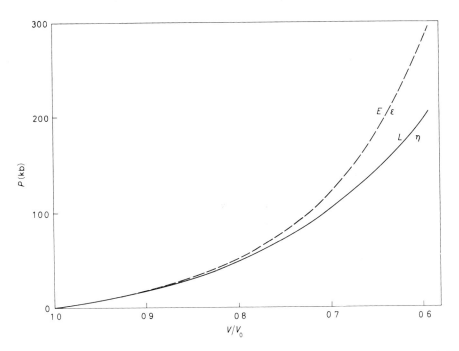

Fig. 2. – Comparative curves of the Lagrangian and Eulerian formulations of finite strain theory, after equations (12) and (17), respectively. The data [6] tend to confirm the Lagrangian theory. $--$ Eulerian formulation ($\varepsilon$);  ———— Lagrangian formulation ($\eta$).

It has been argued [5b] that the series (17) converges faster than (12). The original argument concerned only $A_3$ and $B_3$: since most materials have values of $K_0'$ between 3.5 and 6.5, $B_3$ is generally much smaller than $A_3$ (cf. eqs. (11c), (16c)). Now that we have better data, we know further that, for CsI for example, $B_4$ is much smaller than $A_4$. These comparisons appear to be coincidental, as the difference $A_i - B_i$ is essentially a geometrical artifact of the definitions (6) and (13), and has nothing to do with the atomic forces which *determine* $K_0'$, $K_0''$, etc. The extension of such comparisons to conclusions on the neglected terms, or on the validity or usefulness of (12) or (17) seems quite hazardous. If one wishes, for some reason, to *assume* $K_0' = 4$, he may do so *either* by setting $B_3 = 0$ or $A_3 = -18 K_0 V_0$, it is *another* assumption to choose between (12) and (17).

**2˙3.** *Ambiguity in « Finite-strain theory »*. – One sees that the theory of finite strain, while based on the *exact* eqs. (4), and (9), (15), etc., leads in practice to inexact, truncated series. There are an infinite number of such series, whose predictions diverge, typically *outside* the range of direct experimental evidence, and among which *no* theoretical superiority can be established at this point in the argument. The way out of this dilemma was indicated by KNOPOFF [7], who observed that any choice of a definition of strain is equivalent to the assumption of a model of atomic forces. His example was a simple one; he defined a strain scalar by

$$(18) \qquad\qquad \varphi = \varepsilon + a_2 \varepsilon^2 ,$$

with $a_2$ a free constant. Through an identity, he showed that this leads to the equation of state, correct to second order,

$$(19) \qquad\qquad P = \frac{3}{2 + 2a_2} K_0 \left[ \left( \frac{V}{V_0} \right)^{-(7+4a_2)/3} - \left( \frac{V}{V_0} \right)^{-(5+2a_2)/3} \right] .$$

That is, the constant $a_2$ appears as an *exponent* of $(V/V_0)$. Then $a_2$ may be chosen to reproduce an equation of state derived on atomistic grounds; for example, with $a_2 = -\frac{1}{2}$, (18) becomes

$$(19a) \qquad\qquad P = -3K_0 \left[ \left( \frac{V}{V_0} \right)^{-\frac{5}{3}} - \left( \frac{V}{V_0} \right)^{-\frac{4}{3}} \right] .$$

This was derived earlier by BARDEEN [8], and results from a total energy function containing ground state terms ($\sim r^{-3}$ and $r^{-1}$) and, in addition, a Fermi energy term ($\sim r^{-2}$). Thus, « in a modest way », KNOPOFF showed to what atomistic model a particular formulation of finite strain theory corresponded. This *particular* example *can* be pursued [9], but the *important* point is that, in general, the results of finite strain theory can be believed only to the extent that one believes the model of atomic forces to which they pertain.

Before we turn to these atomistic considerations, it is well to mention one further equation which often appears in the literature. The « Murnaghan equation »,

$$(20) \qquad\qquad P = \frac{K_0}{K_0'} \left[ \left( \frac{V}{V_0} \right)^{K_0'} - 1 \right] ,$$

was used by ANDERSON [10] to indicate that extrapolation of low pressure ultrasonic data far beyond its experimental range yelds predictions that are much more reasonable than one might naïvely expect. The mathematical assumption underlying (20), *i.e.*

$$(21) \qquad\qquad K = K_0 + K_0' P$$

is of the same genre as the assumptions underlying (17), and (17) and the same criticisms apply.

## 3. – Statistical mechanics.

A fundamental result of statistical mechanics is that the free energy of a system is statistically related to the possible energy levels of the system by:

$$(22) \qquad\qquad F = - kT \ln Z$$

where $Z$ is the *partition function* defined by

$$(23) \qquad\qquad Z = \sum_n \exp\left[ -\frac{\varepsilon_n}{kT} \right]$$

with $\varepsilon_n$ a possible energy level. $\varepsilon_n$ contains both static and dynamic contributions; the discussion of the latter in the context of a crystal lattice is called *lattice dynamics*. In a solid where a potential energy function $\Phi$ can be defined [11], it may be assumed to be Taylor-expandable about the equilibrium configuration of the solid:

$$(24) \qquad \Phi([\boldsymbol{R} + \boldsymbol{q}(t)]) = \Phi_0([\boldsymbol{R}]) + \sum_I \Phi_I([\boldsymbol{R}]) q_I(t) + \frac{1}{2} \sum_{IJ} \Phi_{IJ} q_I q_J +$$

$$+ \frac{1}{3!} \sum \Phi_{IJK} q_I q_J q_K + \frac{1}{4!} \sum \Phi_{IJKL} q_I q_J q_K q_L + \cdots$$

where $[\boldsymbol{R}]$ is the equilibrium configuration, and $q_I(t)$ is the instantaneous displacement of a particle from equilibrium. (The summation over $I$ covers all vector components of all particle displacements.) A truncation of the series (24) represents a definite approximation in lattice dynamics; several such will be discussed below. It is customary to maintain the same approximation in all configurations, that is, under all external pressures, and this determines the forms of the coefficients $\Phi_{IJ}$ etc. That is, in the *harmonic* approximation, one assumes that $\Phi_{IJK}([\boldsymbol{R}]) = 0$; since $\partial \Phi/_{IJ} \partial R_K = \Phi_{IJK}$ we must have also that the *derivative* of $\Phi_{IJ}$ is 0 near the equilibrium configuration $[\boldsymbol{R}]$. But since this assumption is made at *all* configurations $[\boldsymbol{R}]$, we further must have the derivative of $\Phi_{IJ}$ zero *everywhere*, *i.e.* $\Phi_{IJ}$ is a constant, independent of volume, in the harmonic approximation. This means in general that (24) is equivalent to a similar expansion, this time about a *fixed* configuration $[\widetilde{\boldsymbol{R}}]$. Further, it implies that the potential energy $\phi_0$ of the static lattice must possess the same sort of truncated Taylor expansion:

$$(25) \qquad \phi_0([\boldsymbol{R}]) = \widetilde{\phi} + \sum \widetilde{\Phi}_I(\boldsymbol{R}-\widetilde{\boldsymbol{R}})_I + \frac{1}{2}\sum \widetilde{\Phi}_{IJ}(\boldsymbol{R}-\widetilde{\boldsymbol{R}})_I(\boldsymbol{R}-\widetilde{\boldsymbol{R}})_J +$$

$$+ \frac{1}{3!}\sum \widetilde{\Phi}_{IJK}(\boldsymbol{R}-\widetilde{\boldsymbol{R}})_I(\boldsymbol{R}-\widetilde{\boldsymbol{R}})_J(\boldsymbol{R}-\widetilde{\boldsymbol{R}})_K +$$

$$+ \frac{1}{4!}\sum \widetilde{\Phi}_{IJKL}(\boldsymbol{R}-\widetilde{\boldsymbol{R}})_I(\boldsymbol{R}-\widetilde{\boldsymbol{R}})_J(\boldsymbol{R}-\widetilde{\boldsymbol{R}})_K(\boldsymbol{R}-\widetilde{\boldsymbol{R}})_L + \ldots$$

where the $\widetilde{\Phi}_{IJ}$ etc. are *constants*.

The definition of the fixed configuration $[\widetilde{\boldsymbol{R}}]$ is arbitrary, although a clever choice may maximize the range of validity of the approximation. It is customary to choose $[\widetilde{\boldsymbol{R}}]$ as that configuration which minimizes $\phi_0$; hence $\widetilde{\Phi}_I$ is zero, and the reference state $[\widetilde{\boldsymbol{R}}]$ so defined is called the *rest* state. It is, of course, somewhat smaller than that observed at the « zero » state defined by $P = 0$, $T = T^0$, since the latter shows the effect of some thermal expansion.

We now consider the important approximations of lattice dynamics.

3˙1. *The static approximation.* – This is, of course, the easiest in principle, and yet one must admit that little progress has been made on this problem, despite the efforts of many good physicists. The work may be divided into two groups, the first of which is characterized by attempts to calculate the force parameters of (25) from first principles. As an indication of the present state of this group, the student is referred to the recent work of LIBERMAN [12], who by solving Schrödinger's equation iteratively to converge upon a self-consistent solution for the electronic wave functions, has calculated the equation of state of simple alkali metals. Typically, he has been able to predict the rest volume, $V$, within about 10 %; his solutions appear to become even more reasonable at high pressure. Direct comparison with experiment is difficult, because the experiments reflect the reality of thermal vibration. Geophysically, such calculations on simple metals have their greatest application to the core, and LIBERMAN has not yet completed his calculations for iron, silicon, etc. The extension to the oxides of the solid mantle does not appear to be on the near horizon. KNOPOFF has discussed the results of the more *ad hoc* TFD model of atoms; it appears somewhat less promising than the self-consistent field technique. On a level of sophistication between these two, the work of BARDEEN [8] has been mentioned.

The second group of work in the static approximation starts with a recognition and avoidance of the problems encountered in the first group. For example, BORN and his collaborators invented the « *n-m* law » which assumed the lattice potential to be the sum of central, two-body terms

$$\phi = \frac{a}{r^m} + \frac{b}{r^n} \,.$$

The first term is attractive, the second repulsive, hence $m < n$. In certain cases, $m$ is « known », *i.e.* it is $m = 1$ in ionic crystals, $m = 6$ in Van der Waals crystals. The other constants $a$, $b$, $n$, may be evaluated in terms of boundary conditions as in Sect. **2**, and $n$ generally turns out to be between 9 and 12. A considerable body of work has been done with this idea, or modifications of it, but since one has little confidence in such « laws », their results inspire little confidence either. However, arguments based on such simple ideas may give valuable physical insight into complex situations; an excellent example is the recent work of ANDERSON [13] and colleagues.

3'2. *The harmonic approximation.* – This is the first *quasi*-realistic approximation, and it has great utility. The statement of the approximation is that $\Phi_{IJK} = 0$ in (24); this leads to $\tilde{\Phi}_{IJK} = 0$ in (25). The total instantaneous energy of the system is then

$$(26) \qquad \mathcal{H}([\mathbf{R} + \mathbf{q}], t) = \frac{1}{2} \sum_{IJ} \Phi_{IJ} q_I q_J + \sum_I \frac{M_I \dot{q}_I^2}{2} \,,$$

where the second term is the kinetic energy. The equations of motion are

$$(27) \qquad M_I \ddot{q}_I = \sum_J \Phi_{IJ} q_J \,,$$

*i.e.* they are some $10^{23}$ *coupled* equations. The coupling is avoided by transforming from the $q_I$ to *normal co-ordinates* $a_I$ by the transformation

$$(28) \qquad q_I = Q_{I\mu} a_\mu \,,$$

where the matrix $Q_{I\mu}$ is defined to possess the property

$$(29) \qquad \sum_{IJ} \Phi_{IJ} Q_{I\mu} Q_{J\nu} = \omega_\mu^2 \delta_{\mu\nu}$$

of diagonalizing the potential energy.

Substituting (28) in (26) and proving that $\mathbf{Q}$ is orthogonal leads to the equations of motion:

$$(30) \qquad M_\mu \ddot{a}_\mu = \omega_\mu^2 a_\mu \,.$$

The equations are uncoupled; further they describe simple harmonic motion. *The equations of motion of a harmonic solid are equivalent to those describing an imaginary set of independent harmonic oscillators, whose co-ordinates are the $a_\mu$.*

We now form the partition function $Z$, eq. (23). We write $F = \sum f_\mu$, with $f_\mu = -kT \ln Z_\mu$ being the free energy of the $\mu$-th mode, and $Z_\mu$ its partition function. Then

$$(31) \qquad Z_\mu = \sum_n \exp\left[-\varepsilon_n^{(\mu)}/kT\right].$$

From the quantum mechanics of a harmonic oscillator, we have, for the $n$-th energy level of the $\mu$-th normal mode,

$$(32) \qquad \varepsilon_n^{(\mu)} = \varphi_\mu + (\tfrac{1}{2} + n)\hbar\omega_\mu ; \qquad\qquad n = 0, 1 \ldots ,$$

where $\varphi_\mu$ is the energy of assembling the static oscillator into the solid lattice. Then, using a well-known mathematical identity to change the summation into a closed expression, we have

$$(33) \qquad Z_\mu = \frac{\exp\left[-(\varphi_\mu + \tfrac{1}{2}\hbar\omega_\mu)/kT\right]}{1 - \exp\left[-\hbar\omega_\mu/kT\right]}$$

and, upon summing, over all modes we have for the free energy:

$$(34) \qquad F = \phi_0 + \sum_\mu \left(\tfrac{1}{2}\hbar\omega_\mu + kT \ln\left(1 - \exp\left[-\hbar\omega_\mu/kT\right]\right)\right),$$

$\phi_0$ is, of course, $\sum \varphi_\mu$.

The problem is now reduced to determining the static potential energy $\phi_0$ and the eigenfrequency spectrum $\omega_\mu$. It is, strictly speaking, inconsistent to graft onto the vibrational part of the harmonic approximation an expression for $\phi_0$ from the static approximation. One should strictly use only the original expression (25), as truncated at the second order:

$$(35) \qquad \phi_0 = \tilde{\phi} + \tfrac{1}{2} \sum \tilde{\Phi}_{IJ}(\boldsymbol{R} - \tilde{\boldsymbol{R}})_I (\boldsymbol{R} - \tilde{\boldsymbol{R}})_J .$$

This may be expressed in terms of the « strain » tensors of Sect. 2 by using (6) or (13) to find, respectively,

$$(36a) \qquad \phi_0 = \tilde{\phi} + \tfrac{1}{2} \sum_{IJ} \tilde{\Phi}_{IJ} \tilde{R}_K \tilde{R}_L \, e_{IK} e_{JL} ,$$

or

$$(36b) \qquad \phi_0 = \tilde{\phi} + \tfrac{1}{2} \sum_{IJ} \tilde{\Phi}_{IJ} R_K R_L f_{IK} f_{JL}$$

in terms of the Lagrangian « strain » tensor $e$ and the Eulerian $f$ respectively. It is immediately clear that the coefficients of $e_{IK}e_{JL}$ in (36a) are *constants,* independent of configuration, while the coefficients of $f_{IK}f_{JL}$ in (36b) *do* depend on the configuration $[R]$, and hence upon the volume. It is seen that the basic concept of lattice dynamics, the description of the energy as a Taylor expansion in displacements rules out the possibility of describing the solid in Eulerian terms. The series (25) can never be equivalent to a polynomial of Eulerian strain tensors with constant coefficients. Thus the model of atomic forces (24), (25) of lattice dynamics is here identified with the Lagrangian finite « strain » definitions $e$ of eq. (6).

We therefore concentrate on the Lagrangian expression (36a). Because of the rotational invariance of $\phi_0$, (36a) is equivalent to a second-order expansion in $\eta$, which we may write as:

$$(37) \qquad \phi_0 = \tilde{\phi} + \tfrac{1}{2}\tilde{V}\sum_{ijkl}\tilde{C}_{ijkl}\eta_{ij}\eta_{kl}\ .$$

Here, each index is summed only over the three directions. The relation of the $\tilde{C}_{ijkl}$ to the $\tilde{\Phi}_{IJ}\tilde{R}_K\tilde{R}_L$ is straightforward, but immaterial. In our simple case of high symmetry, it may be further simplified to

$$(38) \qquad \phi_0 = \tilde{\phi} + \frac{9}{2}\tilde{K}\eta^2\ ,$$

where

$$(39) \qquad \eta = \frac{1}{2}\left[\left(\frac{V}{\tilde{V}}\right)^{\frac{2}{3}} - 1\right]$$

similar to (10). This is not a realistic potential, and it can be shown to be grossly inadequate, even at moderate pressure. But the harmonic approximation *itself* is unrealistic and inadequate, being mainly a training ground for more accurate treatments. Historically, inconsistent expressions, such as an *n-m* law, have been often utilized for $\phi_0$, and perhaps this procedure is in fact appropriate at this level of sophistication.

The eigenfrequencies, $\omega_\mu$, are through (29) quantities of the same order (the second) as $\Phi_{IJ}$ and are hence constant in volume. Thus, the thermal equation of state, from (4), (34), and (38) is

$$(40) \qquad P = -\frac{\mathrm{d}\phi_0}{\mathrm{d}V} = -3\tilde{K}\left(\frac{V}{\tilde{V}}\right)^{-\frac{1}{3}}\eta$$

and the pressure is seen to contain no vibrational effects. The frequencies $\omega_\mu$

enter, in a nontrivial way, only into the *caloric* equation of state. From (3), (34):

$$(41) \qquad S = -k \sum \ln\left(1 - \exp\left[-\hbar\omega_\mu/kT\right]\right) - k \sum \frac{\hbar\omega_\mu/kT}{\exp\left[\hbar\omega_\mu/kT\right] - 1}$$

and

$$(42) \qquad C_V = -T\left(\frac{\partial S}{\partial T}\right)_V = k \sum \frac{(\hbar\omega_\mu/kT)^2}{(\exp\left[\hbar\omega_\mu/kT\right] - 1)^2} \exp\left[\hbar\omega_\mu/kT\right],$$

The details of the frequency spectrum are not crucial to evaluating the sum (42). In fact, one sees immediately that independently of these details, at high temperatures $(T > \hbar\omega_{max}/k \equiv \theta)$ the molar specific heat becomes simply $3sN_0k/Z$ where $sN_0/Z$ is the number of particles, and at low temperatures $(T \ll \theta)$, it vanishes. The trajectory of the specific heat curve between these limits is given surprisingly accurately by Einstein's [14] simple assumption that all the frequencies are equal. A more realistic assumption was made by DEBYE [15], who derived an analogy between the vibrating atomic solid and a perfect elastic continuum. The continuum propagates elastic waves with constant velocities $V = L\omega/2\pi$ over a large range of frequencies. The maximum frequency corresponds to the minimum wave-length, which must be of the order of the atomic spacing, $l$; hence $\omega_{max} \approx 2\pi V/l$. It remains to enumerate the different sorts of waves, and make a properly weighted sum of the corresponding maximum frequencies in order to arrive at an « effective » temperature $\theta_{eff}$, above which $C_v$ may be considered constant. Specifically, DEBYE considered only the existence of longitudinal and transverse acoustic waves, and arrived at the formula:

$$(43) \qquad \theta_D \equiv \frac{\hbar}{k}\,\omega_D \equiv \frac{2\pi\hbar}{k}\left(\frac{3}{4\pi}\right)^{\frac{1}{3}}\left(\frac{3sN_0}{ZV}\right)^{\frac{1}{3}}\left[\frac{2}{V_\perp^3} + \frac{1}{V_\parallel^3}\right]^{-\frac{1}{3}}.$$

Despite the crudity of Debye's model, this equation is of considerable importance, as it defines approximately, in terms of experimental quantities, a characteristic temperature above which the thermal properties of solids are relatively simple. It is a relief to geophysicists to discover that all present models of the temperature distribution of the Earth lie above the curve of $\theta_0$ plotted from (43) with the aid of seismic velocity profiles.

BORN and HUANG [16] (Sect. 4) give a fuller discussion of the frequency models of Einstein and Debye; they also show (Sect. 5) that the actual frequency spectrum must be far more complex than these would suppose. In fact, for a three-dimensional isotropic body, the dispersion diagram appears schematically as in Fig. 3. It is seen that the model of Debye has ignored the existence of a whole class of possible waves, the *optic* branches, which have nothing to do at all with the propagation of sound. Further, it has ignored the phenomenon

of dispersion at the high-frequency end of the acoustic spectrum. It is remarkable, then, that the model is as useful as it is. KNOPOFF has shown in these lectures that for many problems, the full complexity of the vibrational spectrum must be taken into account. Finally, one is reminded that the connection between the eigenfrequencies of vibration and the velocities of acoustic propagation is known *only* in this harmonic approximation (cf. ref. [17], Sects. **8, 20**).

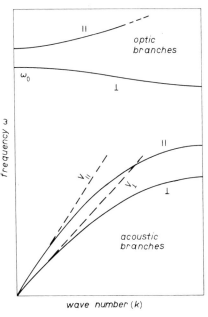

The constants $\tilde{K}$, $\tilde{V}$ appearing in (38), (39) may be evaluated in two ways. If the *microscopic* potential between the various atoms is known or assumed, then the constant $\tilde{K}$, for example, may be calculated directly [25] as a function of the $\Phi_{IK}\tilde{R}_I\tilde{R}_K$. On the other hand, if, as is generally the case, an adequate microscopic potential is not available, then $\tilde{K}$ and $\tilde{V}$ can be evaluated directly from experimental quantities. For example, from (40), at $P = 0$, $\eta = 0$, and from (34), $\tilde{V} = V_0 =$ the volume at zero pressure, room temperature. Likewise from (40), $K = -V(\partial P/\partial V)_T = \frac{1}{3}P + \tilde{K}(V/\tilde{V})^{\frac{4}{3}}$, so that $\tilde{K} = K_0$: (These relations are valid *only* in the harmonic approximation!) As an intermediate case, if a model potential is assumed (such

Fig. 3. – Not indicated in this schematic diagram of the various types of lattice vibrations is either the density of discrete points which together define the «lines» shown, or the change in shape and position of the «lines» with respect to a change in the volume of the solid.

as the two-body *n-m* law) then the adjustable parameters in it may be evaluated by $\tilde{K}$ and $\tilde{V}$, which in turn have been evaluated by the experimental quantities $K_0$, $V_0$. Then the additional parameters which will appear in the higher-order theories may be calculated from these. We turn now to these higher-order theories.

**3˙3.** *The third-order approximation.* – The harmonic approximation is unsatisfactory because it does not include many common experimental results, *e.g.* the thermal expansivity $\alpha$, and $K' = (\partial K/\partial P)_T$.

Figure 4 shows schematically the physical idea behind the necessity for an anharmonic theory to describe the thermal expansivity. The two potential wells show *a*) harmonic and *b*) cubic interaction between two particles. Classically, the particles vibrate in their potential well, with amplitudes determined by their kinetic energy (or equivalently, the temperature) and with a mean separation indicated by the vertical lines. Because of the cubic term, the asymmetric cubic well requires a mean separation that increases with temperature, that is, a thermal expansion. The harmonic well, on the other hand, has the same mean separation at all $T$.

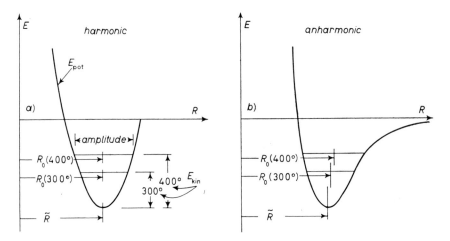

Fig. 4. – The anharmonic lattice possesses a thermal expansion because of its asymmetry. The *mean* separation $R_0(T)$, as well as the amplitude, increases with increasing kinematic energy (*i.e.* increasing $T$). *a*) Harmonic. *b*) Anharmonic.

Mathematically, the anharmonic solid must be treated as a perturbation on the harmonic solid; the definitive paper in this connection is by LEIBFRIED and LUDWIG [17], L&L. The problem is that the equations of motion are now not solvable exactly; the equations of motion (27) contain higher-order non-linear terms. Normal modes can still be defined by (28), (29), but the resulting equations analogous to (30) still contain the anharmonic terms, rendered even more intractable by the transformation. The normal modes still exist, but they describe coupled, anharmonic oscillators. Rather than solve *this* set of equations by perturbation techniques, L&L (Sects. 3, 5) show how they may be bypassed, and the free energy itself may be perturbed. The result is

$$(44) \qquad F = \phi_0(V) + \sum \left[ \tfrac{1}{2}\hbar\omega + kT \ln\left(1 - \exp\left[-\hbar\omega/kT\right]\right) \right] + F_{\text{anh}}(T) \,,$$

(compare with (34)). $\phi_0$ now contains third-order terms, and may be written

$$(45) \qquad \phi_0(V) = \tilde{\phi} + 9K[\tfrac{1}{2}\eta^2 - \tfrac{1}{6}\Gamma\eta^3] \,,$$

(compare with (38)). The second term of (44), the « vibrational contribution », has the same form as in the harmonic approximation (34), only now the $\omega_\mu$

are functions of volume, since $\tilde{\Phi}_{IJK} \neq 0$. $F_{anh}$, the explicit anharmonic contribution, is given by first-order perturbation theory. Because the cubic term of (24) is essentially antisymmetric, in a (by definition symmetric) Bravais lattice, $F_{anh} = 0$. As no physical insight is to be gained by considering more complex lattices (see L+L, Sect. 106) we have the result that the free energy (44) of the third-order approximation is of the same form exactly as that of the harmonic approximation (34). One may call it a *quasi-harmonic* approximation if one is careful to not imply any lack of *rigour* in the treatment, and if one mantains faith with the original statement of the approximation.

The pressure is now given (with (4)) by the « Mie-Grüneisen [18] formula »

$$(46) \qquad P(V, T) = -\frac{d\phi_0}{dV} + \frac{1}{V}\sum_\mu \gamma_\mu \hbar\omega_\mu \left(\frac{1}{2} + \frac{1}{\exp[\hbar\omega/kT] - 1}\right),$$

where

$$(47) \qquad \gamma_\mu \equiv -\frac{V}{\omega_\mu}\frac{d\omega_\mu}{dV}$$

is called the Grüneisen parameter for the $\mu$-th mode. In the high temperature limit, we have, *without further approximation*,

$$(48) \qquad P(V, T) = -\frac{d\phi_0}{dV} + \gamma\frac{3sNkT}{V},$$

where

$$(49) \qquad \gamma \equiv \frac{1}{sN}\sum_\mu \gamma_\mu.$$

This parameter may be evaluated in terms of experimental quantities by operating on (48) with $(\partial/\partial T)_P$ to yield

$$(50) \qquad \gamma = \frac{V\alpha K}{C_V}.$$

This formula for $\gamma$ is strictly valid near $V = \tilde{V} =$ the reference volume (or equivalently near $P = 0$.)

For higher pressures, the formula is inaccurate, simply because $\gamma$ is a third-order quantity (it depends on the terms $\Phi_{IJK}$ of (24)). Its volume-dependence is a *fourth*-order quantity, and is not included in the present treatment. In an attempt to sidestep this fundamental difficulty, several imaginative attemps have been made to define the Grüneisen parameters in terms of theories outside lattice dynamics itself. For example, one may define the derivative $-(V/\omega_0)(\partial\omega_0/\partial V)_T$ of the Debye frequency (see eq. (43)) as an

« acoustic Grüneisen parameter », and hope that the measured volume-dependence of the acoustic velocities will account for $\gamma$ and its volume-dependence. Remarkably, at zero pressure, the « acoustic Grüneisen parameter » thus defined is generally within 25 % of the true $\gamma$, eq. (50). However this sizeable discrepancy indicates that the volume-dependence of the « acoustic Grüneisen parameter » must be even less reliable. KNOPOFF and SHAPIRO [19] have shown lucidly why *all* such « outside » formulations of other « Grüneisen parameters » are equally inadmissable.

The foregoing discussion (begining with (48)) was limited to high temperatures. Below $\theta_0$, the thermal equation of state itself (46) involves the full complexity of the frequency spectrum. A convenient approximation to simplify the equation is to set all the $\gamma_\mu$ of (47) equal. This assumption, due to GRÜNEISEN [18], is certainly false, but it has been very useful in understanding the basic phenomena involved, in the same manner as has the Debye model. The lattice sum in (46) then may be simplified some algebraic operations, leaving the « Mie-Grüneisen equation »:

$$(51) \qquad\qquad P(V, T) = -\frac{d\phi_0}{dV} + \gamma\,\frac{U_s}{V},$$

where $U_s$ is the vibrational contribution to the internal energy:

$$(51a) \qquad\qquad U_s = \sum_\mu \hbar\omega_\mu \left(\frac{1}{2} + \frac{1}{\exp\left[\hbar\omega_\mu/kT\right]-1}\right)$$

($s$ stands for *schwingen*.) The Grüneisen parameter is given, as before, by (50).

This low- and moderate-temperature equation of state (51), approaching (48), of course, at high $T$, is of direct interest to geophysics mainly because the laboratory experiments, in terms of which the composition, etc., of the Earth must be interpreted, are generally performed in the moderate-temperature range, and hence must be corrected for thermal as well as pressure effects.

It is seen that the third-order approximation in lattice dynamics is hardly more difficult, for the study of the equation of state, than is the harmonic approximation. Its greatest shortcoming is the following; the incompressibility, given from (51) by

$$(52) \qquad\qquad K = -V\left(\frac{\partial P}{\partial V}\right)_T = V\,\frac{d^2\phi_0}{dV^2}$$

is seen to be independent of $T$. (The thermal term of (51) disappears in this derivative because it is a third-order quantity, and hence constant in volume.) Since a $T$-dependent $K$ is a common observation, we want a theory which contains this effect; we want the next higher approximation.

3`4. *The fourth-order anharmonic theory.* – We return to eq. (24), as written complete to fourth order, and attempt to form a perturbed free energy function similar to (44). The credit for this perturbation solution as well as for the recognition of the implications of the result, goes to L&L [17]. They showed that the first-order perturbation of the harmonic free energy leads now to a *finite* contribution to $F_{anh}$, as the former symmetry considerations do not eliminate the symmetric quartic terms. Further they showed that the contribution from the second-order perturbation must be of the same order, and they wrote down $F_{anh}$ explicitly. It is a formidable tangle of mode frequencies and energies, but it need not concern us excessively. This is because L&L showed also that, *in the fourth-order approximation*, $F_{anh}$ is a function of temperature only, and hence is irrelevant to the *thermal* equation of state. In forming the pressure with (4), it disappears, and the resulting equation of state is (46) as before. Thus in the fourth-order anharmonic theory, the *thermal* equation of state has still a *quasi*-harmonic appearance, and only the *caloric* equation of state involves the full complexities of the anharmonicity. This shows the extent of the apparently simple division first made between eqs. (3) and (4).

Having established this result, the formal differences between the third-order and fourth-order approximations are slight (for the *thermal* equation of state). $\phi_0$ is now a *quartic* polynomial in $\eta$; the $\omega_\mu$ are *quadratic* in $\eta$; they may be written

(53a)
$$\phi_0 = \tilde{V}\tilde{K}\left\{\frac{(-3)^2}{2!}\eta^2 + \frac{(-3)^3}{3!}\Gamma\eta^3 + \frac{(-3)^4}{4!}\lambda\eta^4\right\}$$

(53b)
$$\ln\omega_\mu = \ln\tilde{\omega}_\mu - 3\tilde{\gamma}_{(\mu)}\eta - \frac{3^2}{2!}\lambda_{(\mu)}\eta^2 .$$

It is convenient though not necessary to make also the Grüneisen approximation, $\tilde{\gamma}_\mu = \tilde{\gamma}$, $\lambda_\mu = \lambda$ and then the equation of state appears as [20]

(54)
$$P(V, T) = -3\tilde{K}\left(\frac{V}{\tilde{V}}\right)^{-\frac{1}{3}}\left\{\eta - \frac{3}{2}\Gamma\eta^2 + \frac{3}{2}\Lambda\eta^3 - \right.$$
$$\left. - \frac{\tilde{U}_s}{\tilde{V}\tilde{K}}\left[\frac{1}{3}\tilde{\gamma} + \left(\lambda - \tilde{\gamma}^2\left(1 - \frac{T\tilde{C}_V}{\tilde{U}_s}\right)\right)\eta\right]\right\}.$$

This approximation is the first serious one that we have mentioned (in the sense that it contains all the primary observable effects, in a perfectly consistent way) and it deserves some elaboration.

Detailed discussions of the fourth-order anharmonic theory have been given elsewhere (caloric equation of state [17], thermal equation of state [20], elastic moduli [21]); perhaps a general discussion will suffice here. Equation (54) is the extension to the domain of finite strain of the theory of lattice dynamics.

However, it may be looked upon from other levels with advantage. For example, it is the Mie-Grüneisen equation (51), with the functions $\phi_o(V)$ and $\gamma(V)$ specified explicitly. The first line in (54) is $-\mathrm{d}\phi_o/\mathrm{d}V$; the second line is the thermal contribution, $\gamma U_s/V$. One sees that the M-G equation is true to the fourth order; in the next Section we will see that it is true *only* to the fourth-order. Thus, to be consistent, any formulation of $\phi_0(V)$ and $\gamma(V)$ used in (51) must be identcal to those contained in (54).

From another point of view, (54) is « simply an extrapolation formula », and this viewpoint is especially convenient for those who do not convince themselves of its theoretical rigour. The constants $\tilde{V}$, $\tilde{K}$, $\Gamma$, $\Lambda$ etc. appearing in the equation are evaluated from data taken near zero $P$ and room $T$ and « simply extrapolated » to higher $P$ and $T$. (Because of the presence of explicit temperature effects in (54), the constants cannot be evaluated simply by formulae such as (11). An iteration procedure is required, and is described elsewhere [21].) The limitation of this « extrapolation » viewpoint is that it does not face up to the fact that the « extrapolation » is a very long one, and just any extrapolation formula will not do. This « extrapolation formula » (54) has two peculiarities, which were justified earlier. First, the extrapolation is in terms of powers of $\eta$, and not another strain measure, like $\varepsilon$, or $\Delta V/V$, which give quantitatively dfferent results. Second, the development is a *fourth*-order one, as required by the temperature-dependence of $K$.

The constants $\tilde{V}$, $\tilde{K}$, $\Gamma$ are related to the $T$-dependent quantities $V_0$, $K_0$, $K_0'$ respectively. The fourth-order coefficient, $\Lambda$, of the static part, is seen, by analogy with the quantity $A_4$, eq. (11$d$) to involve the second derivative $K_0''$. Since the measurement of $K_0''$ is beyond present experimental techniques for most solids, there has been no experimental motivation for a fourth-order theory to utilize it. The point of view of the anharmonic theory is different: a fourth-order quantity $\Lambda$ is theoretically required; one must *find* a way to evaluate it. There now exists a battery of ways to evaluate $\Lambda$, ranging from extremely accurate, moderate-pressure ultrasonic experiments [20], to high-temperature zero-pressure ultrasonic experiments [21], to dynamic shock-wave experiments [20], to microscopic theoretical models [25]. Its evaluation does not appear to be a great problem.

Two more constants, $\tilde{\gamma}$ and $\lambda$ appear in (54) that have not ocurred before. Their appearance here, as contrasted with their absence in the B-M equation (17), is justified because they describe thermal effects which are not included in (12) or (17). They are related to $\alpha$ and $(\partial K/\partial T)_P$ respectively. They specify the Grüneisen parameter $\gamma(V)$ as a function of volume. It is, using (53$b$) and (39)

$$(55) \qquad \gamma(V) = -\frac{\mathrm{d}\ln\omega}{\mathrm{d}\ln V} = \left(\frac{V}{\tilde{V}}\right)^{\frac{2}{3}}(\tilde{\gamma} + 3\lambda\eta).$$

This expression is rigorous, and not subject to criticisms [19] of estimates of $\gamma(V)$ which come from outside lattice dynamics proper. This is because in (55), no assumptions were made concerning the details of the eigenfrequency spectrum; the formula follows from the basic statement of the fourth-order approximation. The possibility of the existence of a « Grüneisen parameter » in the next higher (fifth-order) approximation is briefly discussed in the next Section. Table I lists the constants of some minerals of geophysical interest.

TABLE I.

|  | NaCl [a] | MgO [a] | Garnet [b] | Spinel [c] | Olivine [d] |
|---|---|---|---|---|---|
| $\tilde{\theta}_0$ | 326° | 998° | 765° | 907° | 760° |
| $V_2/\tilde{V}$ | 1.0373 | 1.0215 | 1.0111 | 1.0138 | 1.0123 |
| $\tilde{K}$, kb | 285.5 | 1734.0 | 1873.8 | 2086.0 | 1370.0 |
| $\tilde{\gamma}$ | 1.566 | 1.75 | 1.149 | 1.371 | 1.042 |
| $\lambda$ | 0.559 | 1.36 | 1.56 | 1.181 | 1.24 |
| $\Gamma$ | 5.09 | 4.51 | 5.26 | 3.700 | 4.99 |
| $\Lambda$ | 23.0 | 22.0 | 20.0 | 19.9 | 20.0 |
| $\tilde{C}_{11}$ | 613.9 | 3351.0 | 3234.6 | 3119.4 |  |
| $\tilde{C}_{12}$ | 121.3 | 924.0 | 1187.8 | 1570.6 |  |
| $\tilde{C}_{44}$ | 138.7 | 1634.0 | 978.5 | 1631.8 | 857.0 |
| $\lambda_{11}$ | 2.330 | 3.87 | 2.131 | 2.032 |  |
| $\lambda_{12}$ | −0.327 | 0.01 | 1.284 | 0.757 |  |
| $\lambda_{44}$ | −0.356 | 0.20 | 1.821 | 0.522 | 1.42 |
| $\Gamma_{11}$ | 13.19 | 11.35 | 8.70 | 5.829 |  |
| $\Gamma_{12}$ | 1.05 | 1.14 | 3.53 | 2.659 |  |
| $\Gamma_{44}$ | 1.35 | 2.30 | 2.36 | 2.081 | 2.86 |
| $\Lambda_{11}$ | 67.0 | 58.9 | 28.0 | 28.9 |  |
| $\Lambda_{12}$ | 1.0 | 3.8 | 16.0 | 15.5 |  |
| $\Lambda_{44}$ | 2.8 | 3.8 | 5.0 | 8.4 | 7.0 |

(a) Cf. THOMSEN [21].
(b) After data of N. SOGA: Journ. Geophys. Res., 72 (16), 4227 (1967).
(c) After preliminary data of R. C. O'CONNELL: private communication.
(d) After VHR average of single-crystal data by M. KUMARAWA and O. L. ANDERSON: Journ. Geophys. Res., 74 (25), 5961 (1969).

3'5. *The fifth-order approximation.* – This one has never been done properly. But we can see some of its features immediately. Within the present framework, the fifth-order free-energy will be (compare (44)) [25]

$$(56a) \qquad F(U,\, T) = \phi_0(V) + \sum_\mu \left( \tfrac{1}{2} \hbar \omega_\mu + kT \ln \left( 1 - \exp\left[ -\hbar \omega_\mu / kT \right] \right) \right) +$$
$$+ F_{\text{anh}}(V,\, T) + F_{\text{h.o.}}(T) \,.$$

The perturbation term $F_{\text{higher order}}$ is seen, even without doing the calculation, to depend on $T$ only, and hence to be irrelevant to the thermal equation of state. The term $F_{\text{anh}}$ is of the *form* determined by L&L, but now containing the appropriate volume-dependence. Because $F_{\text{anh}}(V, T)$ does not vanish with the differentiation $(\partial/\partial V)_T$, the equation of state will be « post Mie-Grüneisen »; it will contain an additive term which, at high temperature, will be proportional to $T^2$:

$$(56b) \qquad P(V, T) = -\frac{d\phi_0}{dV} + \gamma(V)\frac{3sNkT}{V} - \tau\frac{(3sNkT)^2}{3\tilde{V}^2\tilde{K}}\left(\frac{V}{\tilde{V}}\right)^{-\frac{1}{3}}.$$

Thus, the M-G equation (48) is valid *only* to fourth order; it *must* be specified by the fourth-order anharmonic theory [54]. However, it is clear that the post M-G equation (56b) may be written as

$$(56c) \qquad P(V, T) = -\frac{d\phi_0}{dV} + \gamma^*(V, T)\frac{3sNkT}{V},$$

where

$$(56d) \qquad \gamma^*(V, T) = \gamma(V) + \tau\left(\frac{V}{\tilde{V}}\right)^{\frac{2}{3}}\frac{sNkT}{\tilde{V}\tilde{K}}.$$

Then the fifth-order anharmonic theory assumes a « *pseudoquasi*-harmonic » appearance. Perhaps it is best to avoid such descriptive names and refer to any approximation in lattice dynamics by its appropriate number.

It should be noted that this brief discussion has begged the basic question of the validity of the assumption of the existence of a potential energy function. This assumption, due to BORN and OPPENHEIMER, is strictly valid *only* to fourth order [16], and this problem must be confronted by any proper fifth-order theory. That is, a Grüneisen parameter which is temperature-dependent at high $T$ is a contradiction in terms, as we presently understand the problem, and should be avoided, if possible. Fortunately, there is little or no experimental motivation for a fifth-order theory. The fourth-order theory appears to be quite adequate for most purposes.

The anharmonic theory of lattice vibrations outlined in this and the preceding Sections shows how, for strict consistency with the treatment of *vibrational* effects, the static term is *forced* to appear as in (53a). This is curious, since the vibrational term is usually much smaller than the static term. In fact, one should *like* to formulate the theory so that the smaller vibrational term is forced into consistency with the larger static term. Unfortunately, this is impossible, as we know only one way to treat the vibrational effects, *i.e.* that described here.

In fact, such a strict consistency may not be desirable. In the absence of all vibrations, the static term still persists, and its best description is not necessarily that given in (25). In fact, it may be that, for a simple solid, a static energy $\phi_0$ may be deduced from fundamental considerations with such rigour as to give one confidence in its predictions. (Of course, each material will require its own unique $\phi_0$, and the concept of « an equation of state for solids » must be abandoned.) In such a case, this « fundamental » $\phi_0$ might be used instead of (53a) for the static term, and only the vibrational term would need to be treated by the present expansion methods. That is, (53b) may be used with the fundamental $\phi_0$, and the resulting inconsistency can be simply ignored. (Presumably, in such a case, the constants $\gamma_{(\mu)}$ and $\lambda_{(\mu)}$ will be evaluated theoretically, rather than experimentally.

In the Earth, this approach is of some interest, mainly for the lower mantle. The closely-packed oxides stable there are at once the most accessible theoretically, and the least accessible experimentally of all the solids in the mantle. Their properties will be the subject of serious thought in the coming decade. The qualities required of a fundamental theory adequate to replace the fourth-order anharmonic theory are high: it must include noncentral forces, distant-neighbor interaction, and complex chemical composition. Such a theory, unfortunately does not exist at present.

**3˙6.** *A word on experiments.* – It is worth-while to mention some of the means of generating high pressures in a modern laboratory, and some of the difficulties and limitations encountered [43]. The classic piston-cylinder apparatus is limited to pressures of the order of 60 kb by friction between piston and cylinder wall, and by the build-up of large tangential tensional stresses in the cylinder, which eventually can cause failure. Higher pressures are generated by other types of apparatus which are designed to avoid these problems. The friction problem is avoided by allowing the pressure-transmitting medium to squeeze out between piston and cylinder. (This naturally requires a solid rather than liquid medium; the limiting pressure is then determined by the flow properties of this solid.) The tangential stress problem is avoided by clever geometries, typically involving cutting the cylinder so that one has in reality a number of anvils which advance on the sample. In this way, pressures of the order of 300-500 kb can be reached. In general, one can reach high pressures only with a small sample, because the applied force is limited.

In these high-pressure cells, the stress is never really hydrostatic because of the shear strength of the solid medium. Aside from this, the magnitude of the pressure is not well known, as the pressure on the sample is related in an indirect and/or unknown way to the force applied externally. Therefore the pressure must be calibrated indepedently, generally in one of two ways: 1) A small bit of a material with a « known » phase change may be included in the cell; when the phase change is observed, that « fixed point » serves to calibrate the apparatus. Of course, the « known » pressure of this « fixed point » is known only through prior inference in another apparatus calibrated indirectly through similar means, and the basic problem of pressure accuracy still remains. 2) Alternatively, the volume change of a material with a « known » equation of state may be monitored continually, along with that of the unknown. Again, this calibration obviously depends on prior inference. In both these inferential problems intrinsic to static high-$P$ experiments, the aid of a rigorous theory such as the fourth-order anharmonic theory, is of obvious importance.

On the other hand, in a modern *dynamic* high-pressure apparatus, other problems are encountered. One is not sure that the region behind the shock front is in thermodynamic equilibrium, nor under pure hydrostatic pressure; indeed the reverse has been argued [23, 24]. Even if these questions are favorably resolved and a Hugoniot curve $P(V)$ is determined, the contribution of the temperature to this curve is still unknown. An equation of state must be assumed in order to separate the termal and compressional effects; the M-G equation (51) is generally chosen. The interpretation of the Hugoniot thus depends on knowledge of the Grüneisen parameter; since most have been interpreted in terms of fallacious formulations of $\gamma(V)$, the conclusions are to that extent untrustworthy. Again the aid of a rigourous theory is of great importance.

Thus one sees that the possibilities for experimental verification of a theory are limited by the need for theoretical support of the experiments. In a situation like this, internal consistency between various experiments, all interpreted with the same theory, is the most one can hope for. The fourth-order anharmonic theory starts from a solid base of verified predictions of variations, and its extension to finite strain now rests on a narrow, but apparently solid base of internally consistent high-$P$ experiments. It should be accurate enough for most geophysical applications.

## 4. – Elasticity under pressure.

The elastic moduli of a solid are the parameters which describe the (infinitesimal) elastic response of the solid to infinitesimal changes in stress. No reference was made in that statement to the magnitude of the initial stress; in fact elastic moduli can be defined at *any* initial stress. We consider states of stress and strain which at first are completely general, and later specialize to the physically relevant case of infinitesimal changes from an initial configuration under finite pressure.

The *stress-strain* relation which generalizes the *equation of state* (4) is [17]:

$$(57) \qquad\qquad T_{ij}^{*} = + \frac{1}{V}\left(\frac{\partial F}{\partial u_{ij}^{*}}\right)_{T},$$

where $\overset{*}{u}_{ij}$ measures the (infinitesimal) difference of the deformed configuration $R_i$ from the « initial » configuration $X_i$:

(58) $$R_i - X_i \equiv u_i \equiv u_{ij}X_j$$

and

$$\boldsymbol{u}^* = \tfrac{1}{2}(\boldsymbol{u} + \boldsymbol{u}^\dagger) .$$

This should be familiar already from infinitesimal elasticity; to make the extension to finite strain, one only needs to introduce the concept of another, « natural » configuration differing from $[\boldsymbol{X}]$ by a finite transformation. This has been done already (eq. (6) for the Lagrangian « strain ») and it only remains to relate the differential $\partial \boldsymbol{u}^*$ to the differentials $\partial \boldsymbol{e}$ and $\partial \boldsymbol{\eta}$ of the finite strain tensors. This is [17, 25]

(59) $$\frac{\partial}{\partial u_{ij}^*} = (\mathbf{1} + \boldsymbol{e})_{im}(\mathbf{1} + \boldsymbol{e}^\dagger)_{nj} \frac{\partial}{\partial \eta_{mn}} .$$

Using (59), eq. (57) becomes:

(57a) $$\boldsymbol{T} = \frac{1}{V}(\mathbf{1} + \boldsymbol{e})\left(\frac{\partial F}{\partial \boldsymbol{\eta}}\right)_T (\mathbf{1} + \boldsymbol{e}^\dagger) .$$

With this, one can discuss the stresses accompanying the passage of an elastic wave. The linearized equation of motion is [21]

(60) $$\varrho([\boldsymbol{X}]) \, \ddot{u}_i = c_{ijkl} \frac{\partial^2 u_k}{\partial X_j \partial X_l} ,$$

where the isothermal elastic modulus is:

(61) $$c_{ijkl} = \left(\frac{\partial T_{ij}}{\partial u_{kl}}\right)_T \bigg|_{U=0} .$$

Because of the form of the wave-equation (60), only the symmetric sum $(c_{ijkl} + c_{ilkj})$ can be determined acoustically. Hence one can also define other elastic coefficients $\boldsymbol{C} \equiv \boldsymbol{c} + \boldsymbol{D}$ where $\boldsymbol{D}$ is limited *only* by the antisymmetry condition $D_{ijkl} = -D_{ilkj}$. Several such definitions are useful; because the elastic moduli in (61) appear in equation of state studies as well as in acoustic propagation, they have been called « effective elastic coefficients » [11, 26]. Operating with (59) and the corresponding

(62) $$\frac{\partial}{\partial \boldsymbol{u}} = (\mathbf{1} + \boldsymbol{e}) \frac{\partial}{\partial \boldsymbol{e}}$$

upon the stress (57a), the elastic moduli (61) become

$$(63) \qquad c_{ijkl} = \frac{1}{V}(1+e)_{im}(1+e)_{kr}\left(\frac{\partial^2 F}{\partial\eta_{mn}\,\partial\eta_{rs}}\right)_T (1+e^\dagger)_{sl}(1+e^\dagger)_{rj} -$$
$$- P[-\delta_{ij}\delta_{kl} + \delta_{lj}\,\delta_{ki} + \delta_{il}\delta_{kj}].$$

Here the equilibrium stress $T_{ij}(\boldsymbol{u}=0)=-P\delta_{ij}$ is explicit. It is clear that (63) contains two sorts of terms. The first sort involves $\partial^2 F/\partial\eta^2$ as is expected for the elastic modulus. The second sort contains the *first* derivative $\partial F/\partial\eta$, and appears in (63) as the pressure. It is a result of chain-rule differentiation, and is important only for finite initial stress. Its appearance means that the elastic modulus is not a homogeneous function of $\eta$. In a fourth-order theory, for example, the first term of (63) will be quadratic in the strain, and the second will be cubic. These may *not* be truncated to a common order in $\eta$.

Expression similar to (63) can be written in terms of any « strain » measure [5b, 21, 27]; the remarks above apply to all. When the fourth-order anharmonic theory is used to define $F$, the result is [21]:

$$c_{\alpha\beta}^s = \left(\frac{V}{\tilde{V}}\right)^{\frac13}\left\{\tilde{c}_{\alpha\beta} - 3\tilde{K}\Gamma_{\alpha\beta}\eta + \frac{q}{2}\,\tilde{K}\Lambda_{\alpha\beta}\eta^2 - \frac{\tilde{u}_s}{\tilde{V}}[\lambda_{\alpha\beta} - \tilde{\gamma}^2\delta_\alpha\,\delta_\beta]\right\} - P\delta_\alpha^\beta$$

for the *adiabatic* elastic moduli of a cubic or isotropic solid. (Compare with the equation of state (54).) Here, we have switched to the Voigt notation, $ij \to \alpha$, and $\delta_\alpha^\beta$ is the coefficient of $-P$ in (63). Again, one is referred to the source [21] for a detailed discussion of this equation; only summary remarks are in order here.

The laboratory measurement of the elastic moduli is restricted in practice to pressures generally less than 15 kb, and to temperatures less than 700 °C. Thus the high-$T$ predictions of (64) may be well tested experimentally, while the high-$P$ predictions cannot. Because at high $T$ the vibrational energy $\tilde{U}_s \sim kT$, (64) becomes linear in the temperature, *at constant volume*. However, the experiments are made at constant (zero) pressure, and so to explain them one must account in (64) for the effects of thermal expansion in the $\eta$ terms. This leads to a curvature at high $T$, whose magnitude depends on the size of $\Lambda_{\alpha\beta}$. This parameter can, in principle, be evaluated in terms of $c_{\alpha\beta}''$, but as mentioned before, this second pressure derivative generally cannot be measured. However, on the basis of a simple atomic model, one may estimate $\Lambda_{\alpha\beta}$, and it turns out to be of the proper sign and size to account for the observed curvature in $c_{\alpha\beta}(T)$. Thus a large part of the curvature is seen to be an effect of finite strain, and not an effect of higher-order anharmonicity. One may then argue that the atomic model is *too* simple, that the *entire* curvature is due to finite strain (*i.e.* to the $\Lambda_{\alpha\beta}$ terms), and thereby one may determine $\Lambda_{\alpha\beta}$ accurately.

Using these $\Lambda_{\alpha\beta}$, one may then venture to apply the formula (64) to high pressures. In the absence of direct experiment, the theory may be verified only indirectly. The strongest verification available is the case of NaCl; this will be discussed in Sect. **6**.

## 5. – The low-velocity layer as a $P\text{-}T$ effect.

A layer in the Earth where one or both of the acoustic velocities decreases with increasing depth is called a low-velocity layer (LVL). The existence of an LVL in the mantle under most if not all oceans, and under many if not most continents, is fairly well established seismologically. Since seismic velocities typically increase strongly with pressure, this conclusion, when first presented by GUTENBERG [28], came as something of a surprise. A tentative explanation immediately presented itself: seismic velocities typically *decrease* with increasing temperature; as $T$ and $P$ both increase with depth in the Earth, the two effects on the velocities obviously compete. Their relative success in this competition depends on the $T$ and $P$ gradients, and also upon the physical properties of the mantle rock. Of course, we know neither of these directly, and so must speculate. But limits can be placed on the range of speculation; one immediately notices that the placement of the LVL in the upper (as opposed to the lower) mantle checks with the fact that the thermal gradient is certainly steeper near the surface than below [29].

Having deduced that the $P\text{-}T$ explanation is worth investigating, it still pays to consider other possible explanations. The most prominent argue that the LVL is due to progressive compositional changes [30], solid-solid mineralogy changes [31], or partial melting [32, 33]. These are all quite possible, and the effect is probably due to some combination of all these explanations. We discuss here mainly the $P\text{-}T$ explanation since the others are outside the domain of the equation of state *per se*. It should be noted that the partial melting explanation depends on the magnitude of the temperature, while the $P\text{-}T$ explanation depends on the magnitude of the temperature gradient; these considerations are obviously related, but they are not identical. Finally, it should be noted that other LVL's at lower depths have appeared in the literature [34], and are amenable to the same debate.

In a homogeneous layer, the variation of a velocity $V$ (either $V_{\perp}$ or $V_{\parallel}$) with depth $Z$ is

$$(65) \qquad \frac{\mathrm{d}V}{\mathrm{d}Z} = \left(\frac{\partial V}{\partial P}\right)_{T} \frac{\mathrm{d}P}{\mathrm{d}Z} + \left(\frac{\partial V}{\partial T}\right)_{P} \frac{\mathrm{d}T}{\mathrm{d}Z} .$$

Since the $T$-gradient is a major imponderable in this situation, the debate is generally cast in terms of what $\mathrm{d}T/\mathrm{d}Z$ is required to generate a given $\mathrm{d}V/\mathrm{d}Z$.

The *P*-gradient is given by the hydrostatic condition:

$$(66) \qquad \frac{\mathrm{d}P}{\mathrm{d}Z} = \varrho g$$

and so we can set $\mathrm{d}P/\mathrm{d}Z = 0.32$ kb/km in the upper mantle. The *T*-gradient is, from (65):

$$(67) \qquad \frac{\mathrm{d}T}{\mathrm{d}Z} = \left[ \frac{\mathrm{d}V}{\mathrm{d}Z} - 0.32 \left( \frac{\partial V}{\partial P} \right)_T \right] \bigg/ \left( \frac{\partial V}{\partial T} \right)_P .$$

There is considerable uncertainty on the r.h.s. of (67), because of 1) the lack of uniqueness [35] of the results of inverting seismic data to yield $\mathrm{d}V/\mathrm{d}Z$ and 2) the lack of precise knowledge of the physical properties $(\partial V/\partial P)_T$ and $(\partial V/\partial T)_P$ appropriate to the upper mantle rock. The former uncertainty can be avoided by discussing only the case $\mathrm{d}V/\mathrm{d}Z = 0$, *i.e.* by discussing the « critical *T*-gradient » at which an LVL first appears. This *critical* gradient depends only on rock properties, and can be discussed in terms of the range of properties in silicates and oxides measured in the laboratory. Under laboratory conditions, the critical gradients turn out [36] to be between 1 and 4 °K/km for shear waves, and between 5.5 and 9.5 for compressional waves, for a suite of minerals of importance in the upper mantle.

The lowest values of $(\mathrm{d}T/\mathrm{d}Z)_{\mathrm{crit}}$ for shear waves are for spinel structures; the microscopic reason for these low values is understood qualitatively [37]. They are most important for discussions of a possible LVL in the transition region, between 450 and 600 km.

ANDERSON and SAMMIS [33] have emphasized that a discussion of critical gradients is not sufficient, for many velocity models show LVL's which are quite pronounced, and for which correspondingly higher *T*-gradients are required of the *P-T* explanation. In fact, upon applying eq. (67) to various velocity models, and to acoustic data on (porous) olivine and pyroxene, they concluded that unreasonably high *T*-gradients were indicated, and they suggested that partial melting was required as explanation.

Now much better data on olivine are available, and their conclusions must be modified. Following their own prediction, the new data on (nonporous) olivine weaken their conclusions somewhat. It is likely that when better pyroxene data will be available, this pattern will continue.

Of equal importance is the following observation, that the physical properties $(\partial V/\partial T)_P$ and $(\partial V/\partial P)_T$ in (67) must be evaluated at the *P* and *T* conditions of the LVL. The elastic moduli depend on *T* according to $\tilde{U}_s$ (cf. eq. (64), or ref. [17].) Hence $(\partial C^s_{\alpha\beta}/\partial T)_V$ is proportional to $\tilde{c}_V$. The same is also true of the *T*-derivative at constant *P*, the thermal expansivity, and the *T*-derivatives of the acoustic velocities. Hence, the value of $(\partial V/\partial T)_P$ at high temperature will be related to that measured near 300 °K by the factor $C_V(T)/C_V(300)$. Also, of course the high-$T$ value of $(\partial V/\partial P)_T$ must be used, but it is easy to show, both theoretically and experimentally, that it is very nearly the same as the low-$T$ value. In addition, the physical properties should be evaluated at ambient $P$, *i.e.* some $(30 \div 50)$ kb, but again it is easy to show that this correction is negligible.

The ratio $C_V(T)/C_V(300)$ depends on the Debye temperature of the solid. For a typical silicate, like olivine, $\theta_0 \sim 750°$, and the entire upper mantle is in the classical region $(C_V(T) \approx 6 \text{ cal/mol})$, whereas the laboratory is in the middle region, and $C_V(300) \approx 4.5$. Thus the correction factor is about 4/3. *Therefore all temperature gradients found in the literature, and calculated from (69) or its equivalents with laboratory data at 300 °K should be multiplied by $\sim 3/4$.* This means, for example, that the *critical* gradients [36] under upper mantle conditions are 0.75 to 3 °K/km from shear waves, and 4.3 to 7.1 for compressional waves. Table II presents the same velocity-model calculations as done

TABLE II. – *Temperature gradients required to generate an* LVL *in a uniform upper mantle.*

| Model | | Olivine $\mathrm{d}T/\mathrm{d}Z$ (°K/km) | Pyroxene $\mathrm{d}T/\mathrm{d}Z$ (°K/km) |
|---|---|---|---|
| GUTENBERG [a] | (P) | 10.2 | 11.3 |
| GUTENBERG [a] | (S) | 7.6 | 6.0 |
| BROOKS [b] | (P) | 13.1 | 13.5 |
| BROOKS [b] | (S) | 7.4 | 6.0 |
| JOHNSON [c] | (P) | 15.3 | 15.0 |
| 200202 [d] | (S) | 11.0 | 8.3 |
| Bilby SE [e] | (P) | 18.9 | 18.8 |
| Bilby NE [e] | (P) | 19.7 | 18.0 |
| Shoal fallon NE [e] | (P) | 15.2 | 15.8 |

[a] *Ann. Geofis. Rome*, **12**, 439 (1959).
[b] *Geoph. Mono. 7.*, edited by MAC DONALD and HISASHIKUNO, vol. **2** (1962).
[c] *Journ. Geophys. Res.*, **72**, 6309 (1967).
[d] D. L. ANDERSON and M. SMITH: *Trans. Am. Geoph. Union*, **49**, 283 (1968).
[e] C. B. ARCHAMBEAU *et al.*: *Trans. Am. Geoph. Union*, **49**, 328 (1968).

by ANDERSON and SAMMIS [33], but incorporating the $C_V$ correction, and the new data [38] for olivine. The limits on the $T$-gradient proposed by THOMSEN [39] must be similarly corrected, but his limits on the $\varrho$-gradient are unaffected.

Looking now at Table II, how can we compare these numbers with the actual temperature gradient? All we can do is to compare with estimates based on nonthermodynamic reasoning. Thermal history calculations [40] have suggested that the gradient between 50 and 150 km is $(5 \div 8)$ °K/km, depending on radioactivity concentration and distribution, and on a low estimate $((10 \ (cm)^{-1})$ of the infrared opacity. These studies are of course obsolete, as they did not consider the loss of heat by solid-state convection. (All else being the same, convection would effectively lower these estimates; but of course all would *not* be the same!) Thermal conductivity-productivity calculations [41], adjusted by petrological arguments, show a $T$-gradient decreasing from 13 °K/km at 50 km to 2 °K/km at 150 km under oceans, with a somewhat smaller range under continents. Electrical conductivity arguments [42] lead to greater extremes.

It is seen that the weight of these estimates lies near the average of the gradients listed in Table I for the shear LVL's. It seems clear that the $P$-$T$ explanation is sufficient in principle to account for most of the observed effect. There are strong arguments, however, which indicate that other factors must also be operative.

One would expect that the LVL would begin just under the Moho, where the thermal gradients are supposedly highest. It does not; in general there appears to be a substantial layer ($\sim 40$ km) intervening between the bottom of the crust and the top of the LVL. (This lid, together with the crust, forms the « lithosphere » of global tectonics.) This fact appears to establish that some chemical and or mineralogical changes occur between Moho and LVL. In addition, the beginning and end of the LVL is too abrupt in many models to be explainable by changes in $T$-gradients. Perhaps the wisest posture here is to avoid basing or rejecting a theory upon a detail (*i.e.* a curvature) of a velocity profile which is known to be 1) imprecise to a certain degree, and 2) nonunique to a major degree. Finally, it is clear that the required $T$-gradients of Table II should be the same for $S$-waves and for $P$-waves in some average sense, under the $P$-$T$ explanation. That they are not is possibly a detail which can be explained away by the uncertainties in profiles, and in material properties.

The phenomenon of the LVL must be a mix of the various explanations advanced, a mix which varies from region to region. In all cases, however, the $P$-$T$ explanation does make a contribution, and in some cases it may be the major effect.

## 6. – Crystal stability.

Figures 5 and 6 show the curves of $V_{\parallel}$, and $\varrho$ *vs.* $Z$ for olivine, spinel, and garnet calculated along a proposed geotherm. Also shown is the profile of a model which satisfies all free oscillation and body wave data. Of course none of these curves is unique, especially in regard to details, but they demonstrate nonetheless an important fact about the Earth: the increase in velocity and density in the Earth between 350 and 900 km is much more than can be explained by simple compression of a homogeneous material. The increased density must be due to phase changes or compositional changes or

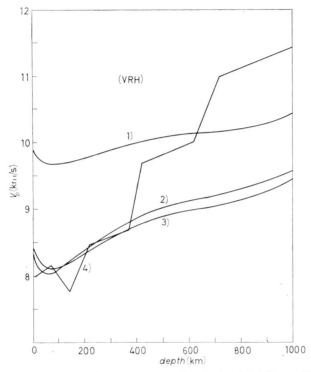

Fig. 5. – The curves for olivine, garnet, and spinel (Voigt-Reuss-Hill average) are calculated [62] using the fourth-order anharmonic theory, and equations similar to (64). The earth model is from JOHNSON [63]. 1) Spinel. 2) Olivine. 3) Garnet. 4) Earth.

both. This was first recognized by BULLEN; suggestions as to the details of the phase changes were given by BIRCH [5c] long before experimental evidence was available.

Now extensive experimental evidence has generally confirmed Birch's speculations and has further revealed a richness of stable high-pressure phases possessed by earthlike materials [45]. The essential transitions are

$$\text{At} \quad \begin{cases} P = (120 \div 150) \text{ kb}, \\ Z = (350 \div 450) \text{ km}, \end{cases} \quad \begin{cases} (215 \div 235) \text{ kb}, \\ (650 \div 700) \text{ km}, \end{cases} \quad \begin{cases} 400 \text{ kb}, \\ 1050 \text{ km}, \end{cases}$$

Olivine      $\rightarrow$ Spinel structure   $\rightarrow Sr_2PbO_4$ structure   $\rightarrow$ $\begin{cases} \text{Phases with} \\ \text{co-ordination } > 6 \,. \end{cases}$
Pyroxenes $\rightarrow$ Garnet  structure $\rightarrow$ ilmenite,  perovskite $\rightarrow$

This presents, of course, an oversimplified picture, but it indicates the essential effects. Recent velocity profiles [46] show features which may be reasonably identified with these transitions; still other profiles [47] reveal an embarrassment of such discontinuities.

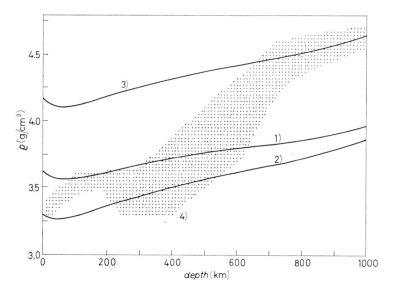

Fig. 6. – The curves for olivine, garnet and spinel (VRH average) are calculated using the fourth-order anharmonic equation of state (54). The stippled region is the envelope of the solutions found by PRESS [53b]. 1) Spinel. 2) Olivine. 3) Garnet. 4) Earth.

The existence of these high-pressure phases has two implications for the equation of state. First, it makes little sense to discuss very high pressure limiting behaviour of an equation of state because invariably before reaching that limit, new phases are encountered, and new equations of state must be established. In the Earth, the largest strains encountered are about $\eta = -0,05$, corresponding to the compression of the close-packed phases between 900 km and the core (or $\eta = -0.1$ corresponding to the compression of the close-packed phases from zero pressure to the core).

Secondly, the phenomenon of high-pressure phases poses a challenge to an equation of state: to predict them and to understand them. The basic criterion for the stability of a given structure is that its Gibbs free energy at given $P$, $T$ be lower than that of all other structures. Now the Helmholtz free-energy of the structure stable at low pressure is given by the fourth-order anharmonic theory; the conversion to the Gibbs function is trivial. The same is true of all other structures, including those stable at high pressure, but the evaluation of the constants is generally impossible. They may be estimated with simplified theories, but as the energy differences are commonly of the order of one or two per cent, more accuracy is generally required than is available.

This problem is avoided by the « necessary » criteria of instability of Born (see ref. [16]). Because the Helmholtz free-energy of a stable structure is a positive definite function of any homogeneous infinitesimal distorsion of the lattice, the elastic moduli of a cubic crystal must *not* satisfy

(68)
$$K < 0 \, ,$$

(68a)
$$c_s = \frac{c_{11} - c_{12}}{2} < 0 \, ,$$

(68b)
$$c_{44} < 0 \, .$$

Further, the same restriction applies to nonhomogeneous infinitesimal distortions, but these are of less interest to geophysics, because they cannot be measured seismically. Similar but more complicated conditions hold for crystals of lower symmetry.

When one of the conditions (70) *is* satisfied, the structure is *necessarily* unstable, and will collapse to another structure, the liquid if necessary. The fact of instability is not *sufficient* to require (70), in fact most instabilities occur without the observance of these conditions. However, it is generally true that as $P$ or $T$ increases towards a critical point, either $c_s$ or $c_{44}$ decreases, thus indicating an incipient instability. In some cases, the necessary conditions may provide a very accurate prediction of the instability. One of the early successes of the fourth-order anharmonic theory was its successful prediction [21] (after the fact!) of a high-pressure instability in NaCl near 287 kb through the vanishing of $c_{44}$.

## 7. – Inferences for the real Earth.

Ideally, one would hope to be able to come from the equation of state, on the basis of seismic data, limited by geochemical and petrographical conclusions, to firm conclusions on the composition and mineralogy of the interior of the Earth. This is at present a real hope, but one which has not yet been realized.

Historically, a number of imaginative schemes for correlating elastic properties of various materials have been proposed, and are still worth mentioning in passing. Birch's law [48] postulated an empirical linear relation ($V_{P0} = a + b\varrho_0$) between zero-pressure compressional velocity and density for a number of rocks and minerals. A correlation of this sort, if precise and unique, would go a long way toward realizing the above mentioned hope (unfortunately, it is neither). The law of corresponding states [49] later recognized the

« coincidence » that *both* $K_0$ and $V_0$ of one oxide compound may be converted into the $K_0$ and $V_0$ of another by changes in $P$ or $T$. The seismic equation of state formalized this coincidence into the rule $V = \overline{M}/\varrho = (1/A)\,\Phi^{-0.323}$. This is essentially (*) the Murnaghan equation of state, burdened with the assumption that it describes a *suite* of rocks at *all* $P$ and $T$, and with the parameters $K_0$, $K_0'$ determined « uniquely » by a least-squares fit of zero-pressure data on these selected rocks. The scatter in the original data is about 4 %; the possibility of extrapolation to high pressures, and especially through phase changes, is unverified. The latter [51] *ad hoc* adjustments of these parameters for various differing classes of materials demonstrate that such simple ideas should be regarded with good humor, while specific conclusions drawn from them should be viewed with extreme suspicion.

We know on broad petrological grounds [54] that the upper mantle is composed of a silicon-oxide matrix with significant amounts of magnesium and iron, somewhat less of aluminium, still less of calcium, and small amounts of many other elements. Acoustic data on rocks and rock-forming minerals, together with an equation of state extrapolating these data to upper mantle conditions, confirm this description, but are not able to greatly refine it [55]. One problem is that the acoustic properties of many candidate materials of various compositions and structures overlap one another, so that a given velocity profile may be interpreted in numerous ways. The unknown $T$-profile exacerbates this situation, as do the imprecision, the nonuniqueness, and the horizontal variability of the seismic results. Another problem is that the relevant acoustic experiments have not yet been done on the high-pressure phase and mineral assemblages which exist in the upper mantle. This will be the most important work in geophysical acoustic in the next five years: to manufacture and measure such samples.

Discrete horizons appearing in seismic profiles offer an especially attractive handle on the composition and mineralogy. Once the seismologists decide where they are with some precision, then specific identifications, as mentioned in the previous Section, can (with some precision) be made with transitions observed in the laboratory. The concentrations of the relatively minor constituents, like Al and Ca, play disproportionately large roles in fixing the critical pressures of these transitions; hence the possibility exists of determining even these minor concentrations.

---

(*) Equation (21) may be written

$$\frac{V}{V_0} = \left(\frac{K}{K_0}\right)^{-1/K_0'} \approx \left(\frac{K^S/\varrho}{K_0^S/\varrho_0}\right)^{-1/(K_0'-1)} \equiv \left(\frac{\Phi}{\Phi_0}\right)^{-1/(K_0'-1)}.$$

$\Phi = K^S/\varrho$ is called the « seismic parameter ».

This relatively promising situation in the upper mantle degenerates below the transition zones. Here the opportunity to duplicate mantle conditions in the laboratory, and to explore the possibilities, does not exist as yet, or in the foreseeable future. (The qualified exception to this statement is in dynamic shock-wave experiments; see Knopoff's lectures.) However, some imaginative efforts have been made to leap this gap. ANDERSON and JORDAN [52] fitted several lower-mantle density solutions to the B-M equation of state [17], chose among the possible fits by comparing to velocity profiles, and concluded that the values of $K_0^s$, $\varrho_0$ thus obtained were much larger than the corresponding parameters of an ordinary material like olivine. This *general* result is correct, of course, although their *specific* results may be easily criticized.

The density solutions they investigated have not been shown to be consistent with free oscillation data; such a study should *start* with a self-consistent set of inverted data ($V_p$, $V_s$, $\varrho$, $P$) [53]. The reliance on the nonrigorous B-M equation, and the approximate treatment of the temperature effects introduce further errors, but these do not damage their *general* result, which may be reached on other grounds as well.

What is the nature of this lower mantle material? In the absence of relevant static experiments, the arguments rely on two indirect approaches. The first approach [45, 46] is through germanate analogues to silicate structures. Since, for example, the Ge olivines undergo phase changes analogous to the Mg, Fe olivines at considerably lower pressures, one may argue that their highest pressure forms correspond also. Of course it develops that there exists a number of possibilities, and the most one can say with confidence about these structures is that they must be quite closely packed. It is probable that they are more dense (at zero pressure) than the average of the $\varrho_0$'s of the component simple oxides MgO, SiO$_2$ stishovite, Al$_2$O$_3$, etc. (One notes that this is barely more than one could conclude on grounds of common sense.) The second approach [57, 58] extrapolates shock-wave data at lower mantle pressures to surface conditions, and discusses the resulting $\varrho_0$, $K_0$ in terms of the crystallographic possibilities indicated by the first approach. Of course the uncertainties in the extrapolation interfere with the conclusions, and hence the details of the results are highly tentative.

Of course the *major* reason for the high $\varrho_0$ and $K_0$ of the lower mantle lies in the closer packing. It is unknown to what extent this effect is supplemented by compositional effects. This discussion is dominated by the possibility of iron enrichment in the lower mantle, and is generally phrased in terms of a possible increase in the mean atomic weight $\bar{M}$. The upper mantle has [54] $\bar{M} = 21.1 \pm 0.16$; values for the lower mantle about one unit above this have been tentatively found [48, 41] but little weight was attached to the difference. Later, differences of greater size were found [52] on similarly marginal grounds, and were tentatively proposed to be real.

RINGWOOD, on the other hand, concluded [56] that the composition was uniform throughout the mantle, and PRESS [53b] suggested an iron enrichment near 1000 km grading towards a net depletion just above the core. No one really understands the properties of matter under lower mantle conditions, and all these suggestions are equally interesting possibilities. The question rests open for the present.

The core presents yet another set of difficulties. This article has been concerned almost exclusively with the solid mantle, as the equation of state of liquids, and that of metals, is quite poorly understood. The core, presenting both these problems, can hardly be discussed without embarrassment; nevertheless, some things can be established. Early arguments [59] based on the existence of iron as well as stony meteorites, proposed that the core was mainly iron, with perhaps a minor percentage of nickel. This is essentially the state of the argument today, although many other proposals [60] have been made concerning the existence and the identity of the alloying elements. Today the consensus indicates an alloy lighter than iron, *e.g.* silicon or silicate. An imaginative proposal [61] that the core-mantle boundary represents an electronic (rather than a crystallographic) phase change to a metallic liquid of the same silicate composition is today discredited by shock wave experiments which indicate no such transition in the relevant pressure range. Statements more detailed than these must wait for further fundamental developments in the theory and practice of the equation of state.

# REFERENCES

[1] Definitive treatises on continuum mechanics and the physics of materials are: *a*) C. TRUESDELL and R. A. TOUPIN: *The classical field theories*, in *Handbuch der Physik*, Vol. **3**/1, edited by S. FLÜGGE (Berlin, 1960); *b*) C. TRUESDELL and W. NOLL: *Nonlinear field theories of mechanics*, in *Handbuch der Physik*, Vol. **3**/3 (Berlin, 1965).

[2] For an entree into this problem, and a review of some pertinent laboratory data, see O. L. ANDERSON *et al.*: *Rev. Geophys.*, **8** (4), 491 (1968).

[3] This subject has been presented in forms far more elaborate, than is necessary for this problem, *viz.* ref. [1 *a*, 1 *b*], and the oft-quoted: *a*) F. D. MURNAGHAN: *Am. Journ. Math.*, **59**, 235 (1937); *b*) F. D. MURNAGHAN: *Finite Deformation of an Elastic Solid* (New York, 1951). One should note that between 1937 and 1951, Murnaghan changed his viewpoint from the « Eulerian » (eq. (17)) to the « Lagrangian » (eq. (12)). The geophysical community has been slow to follow.

[4] The labels « Lagrangian » and « Eulerian » are historically incorrect, and their replacement by the descriptive terms « material » and « spatial », respectively, has been proposed. An infinity of other definitions of finite « strain » is possible

within each of these viewpoints. The Eulerian definitions are only permitted for special problems, which however include the cases of pure hydrostatic pressure, and isotropic bodies (see ref. [1a], Sect. **13-33**; [1b], Sect. **42, 43, 19**).

[5] In an important series of papers, Birch has applied the Eulerian formulation of finite strain to problems of the earth's interior: a) F. BIRCH: *Journ. Appl. Phys.*, **9**, 279 (1938); b) *Phys. Rev.*, **71**, 809 (1947); c) *Journ. Geophys. Res.*, **57**, 227 (1952).

[6] D. E. HAMMOND et al.: *EOS Trans. Am. Geophys. Union*, **51**, 419 (1970). Also: D. E. HAMMOND: *The P-V relationship of CsI up to 250 kb at room temperature*, M.S. Thesis, University of Rochester, Rochester, N. Y. (1970).

[7] L. KNOPOFF: *Journ. Geophys. Res.*, **68**, 2929 (1963).

[8] J. BARDEEN: *Journ. Chem. Phys.*, **6**, 372 (1938); N. H. MARCH: *Physica*, **22**, 311 (1956).

[9] P. BENEVIDES-SOARES: *Ann. de Geophys.*, **24**, 895 (1968).

[10] O. I. ANDERSON: *Phys. Chem. Solids*, **27**, 546 (1966).

[11] For possible limitations on this class of systems (*e.g.* the case of metals), see ref. [16], p. 15, 170, 171, 218, 406, 107; M. TOSI and T. ARAI: *Stability of solids under pressure*, in *Advances in High Pressure Research*, edited by R. S. BRADLEY, Vol. **1** (London, 1966), p. 265.

[12] D. A. LIBERMAN: *Self-consistent field calculations of bulk properties of solids*, paper presented at *Colloque International du Centre National de la Recherche Scientifique sur les Propriétés Physiques des Solides sous Pression, Grenoble, Septembre 1969*. Also, *Phys. Rev. B*, **2**, 244 (1970).

[13] O. L. ANDERSON and R. C. LIEBERMANN: *Phys. Earth Planet. Interiors*, **3**, 61 (1970).

[14] A. EINSTEIN: *Ann. der Phys.*, **22**, 180 (1907).

[15] P. DEBYE: *Ann. der Phys.*, **39**, 789 (1912).

[16] M. BORN and K. HUANG: *Dynamical Theory of Crystal Lattices* (London, 1954).

[17] G. LEIBFRIED and W. LUDWIG: *Theory of anharmonic effects in crystals*, in *Solid State Physics*, edited by F. SEITZ and D. TURNBULL, Vol. **12** (1961), p. 275.

[18] E. GRÜNEISEN: *Ann. der Phys.*, **39**, 257 (1912). More accessible is E. GRÜNEISEN: *Zustand des festen Körpers*, in *Handbuch der Physik*, Vol. **10** (Berlin, 1926), p. 1, available in translation as NASA Republication RE 2-18-59 W.

[19] L. KNOPOFF and J. N. SHAPIRO: *Journ. Geophys. Res.*, **74**, 1435 (1969). One should refer to this paper whenever one encounters a reference in the literature to the « Slater gamma », the « Dugdale-MacDonald gamma », etc., or the statement that the treatment of the variation of the Grüneisen parameter with volume « makes no difference ».

[20] L. THOMSEN: *Journ. Phys. Chem. Solids*, **31**, 2003 (1970).

[21] L. THOMSEN: *The fourth-order anharmonic theory: elasticity and stability*, *Journ. Phys. Chem. Solids*, in press (1971).

[22] J. S. WEAVER et al.: *Calculation of the P-V relation for the BI phase of NaCl up to 300 kb at 25 °C*, paper presented at *National Bureau of Standards Symposium on the Accurate Characterization of the High Pressure Environment* (Gaithersburg, Md., 1968).

[23] D. H. TSAI: *Theory of shock compression of a crystalline solid*, paper presented at *National Bureau of Standards Symposium on the Accurate Characterization of the High Pressure Environment* (Gaithersburg, Md., 1968).

[24] L. V. ALT'SHULER et al.: *Sov. Phys. JETP*, **11**, 766 (1960); M. VAN THIEL and A. KUSUBOV: *Effect of 2024 aluminium alloy strength on high-pressure shock*

*measurements*, paper presented at *National Bureau of Standards Symposium on the Accurate Characterization of the High Pressure Environment* (Gaithersburg, Mb., 1968).

[25] L. THOMSEN: *On the fourth-order anharmonic equation of state of solids*, Ph. D. Thesis, Columbia University (New York, 1968).

[26] R. N. THURSTON: *Journ. Acoust. Soc. Amer.*, **37**, 348 (1965), Erratum, **37**, 1147 1965).

[27]    SAMMIS *et al.*: *Journ. Geophys. Res.*, **75**, 4478 (1970).

[28] B. GUTENBERG: *Zeits. Phys.*, **27**, 111 (1926).

[29] See Lubimova's lectures. Also, one knows that if the measured surface gradients were extended undiminished much into the mantle, the melting curves of silicates would soon be passed.

[30] N. I. CHRISTENSEN: *Tectonophysics*, **6**, 331 (1968).

[31] A. I. RINGWOOD: *Composition and evolution of the upper mantle*, in *The Earth's Crust and Upper Mantle*, edited by P. J. HART, p. 1 (1969). An important book for the next decade.

[32] D. SHIMOZURA: *Journ. Phys. Earth*, **11**, 19 (1963).

[33] D. L. ANDERSON and C. SAMMIS: *Phys. Earth Planet Interiors*, **3**, 41 (1970).

[34] A. B. K. IBRAHIM and O. W. NUTLEY: *Bull. Seis. Soc. Am.*, **57**, 1063 (1967); D. L. ANDERSON and B. R. JULIAN: *Journ. Geophys. Res.*, **74**, 3281 (1969).

[35] See, for example, Press' lectures of ref. [53*b*].

[36] R. C. LIEBERMANN and E. SCHREIBER: *Earth Planet. Sci. Lett.*, **1**, 77 (1969).

[37] O. L. ANDERSON: *Journ. Geophys. Res.*, **73**, 7707 (1968).

[38] M. KUMAZAWA and O. L. ANDERSON: *Journ. Geophys. Res.*, **74**, 5961 (1969).

[39] L. THOMSEN: *Journ. Geophys. Rev.*, **72**, 5649 (1967).

[40] G. J. F. MACDONALD: *Journ. Geophys. Res.*, **69**, 2933 (1964).

[41] S. P. CLARK and A. E. RINGWOOD: *Rev. Geophys.*, **2**, 35 (1964).

[42] D. C. TOZER: *The electrical properties of the earth's interior*, in *Physical Chemistry of Earth*, edited by L. H. AHRENS *et al.*, Vol. **3** (London, 1959), p. 414.

[43] A fuller discussion of static apparatus is given by R. C. NEWTON: *The status and future of high-static pressure geophysical research*, in *Advances in High-Pressure Research*, edited by R. S. BRADLEY, Vol. **1** (London, 1966), p. 195. An insider's explanation of dynamic shock-wave experiments is given by Ahrens in these volume at p. 157.

[44] For a historical account, see K. E. BULLEN: *An Introduction to the Theory of Seismology* (Cambridge, 1963).

[45] For a review of these results, see A. E. RINGWOOD and D. H. GREEN: *Phase transitions*, in *The Earth's Crust and Upper Mantle*, edited by P. J. HART p. 637 (1969).

[46] C. B. ARCHAMBEAU *et al.*: *Journ. Geophys. Res.*, **74**, 5825 (1969).

[47] L. JOHNSON: *Bull. Seis. Soc. Am.*, **59**, 973 (1969); WHITCOMB and D. L. ANDERSON: in preparation.

[48] F. BIRCH: *Geophys. Journ.*, **4**, 295 (1961).

[49] O. L. ANDERSON: *Journ. Geophys. Res.*, **71**, 4963 (1966). A discussion of the possibilities and limitations of simple correlation concepts is given in ref. [2], p. 511.

[50] D. L. ANDERSON: *Geophys. Journ.*, **13**, 9 (1967).

[51] D. L. ANDERSON: *Journ. Geophys. Res.*, **74**, 3857 (1969).

[52] D. L. ANDERSON and T. JORDAN: *Phys. Earth Planet. Interiors*, **3**, 23 (1970).

[53] Such sets have recently been provided by *a*) K. E. BULLEN and R. A. W. HADDON:

*Proc. Nat. Acad. Sci.*, **58**, 846 (1967) (HB$_1$); *b*) E. Press: *Phys. Earth Planet. Interiors*, **5** (1969) (Monte Carlo); *c*) R. A. W. Haddon and K. E. Bullen: *IASPEI Meeting* (Madrid, 1969) (HB$_2$); *d*) D. L. Anderson and T. Jordan: *EOS Trans. Am. Geophys. Union*, **51**, 362 (1970) (a « stable stochastic extension of the pseudoinverse method »).

[54] Major contributors to this consensus are: *a*) A. E. Ringwood: *Mineralogy of the mantle*, in *Advances in Earth Science*, edited by P. M. Hurley (Cambridge, Mass., 1966), p. 357 (pyrolite); *b*) I. G. White: *Earth Planet Sci. Lett.*, **3**, 11 (1967) (peridotites); *c*) W. G. Nelson *et al.*: *Science*, **155**, 1532 (1967) (St. Paul's rocks); *d*) K. Ito and G. Kennedy: *Am. Journ. Sci.*, **265**, 519 (1967) (Kimberlite).

[55] However, for a unified description of some refinements which are as valid as any now current, see D. L. Anderson: *The petrology of the mantle*, in *Min. Soc. Am.*, Spec.papers **3**, 85 (1970); and the lectures of Profs. Press and Jacobs of this volume.

[56] A. E. Ringwood: *Phase transformations in the mantle*, Pubbl. 666, Dept. of Geoph. and Geoch., Australian Nat. Univ., Canberra, 1966.

[57] R. G. McQueen *et al.*: *Journ. Geophys. Res.*, **72**, 4999 (1967).

[58] T. J. Ahrens *et al.*: *Rev. Geophys.*, **7**, 667 (1969).

[59] Attributed as early as 1873 to J. D. Dana by J. A. Jacobs *et al.*: *Physics and Geology* (New York, 1959).

[60] Some high points in this debate were ($M_{iron} = 26$): *a*) W. M. Elsasser: *Science*, **113**, 105 (1951) ($\bar{M} = 29$); *b*) L. Knopoff and R. J. Uffen: *Journ. Geophys. Res.*, **59**, 471 (1954) ($\bar{M} = 22$); *c*) F. Birch: *Journ. Geophys. Res.*, **57**, 227 (1952) ($\bar{M} \leqslant 26$); *d*) S. P. Clark and A. E. Ringwood: *Rev. Geophys.*, **2**, 35 (1964) ($\bar{M} \leqslant 26$).

[61] W. H. Ramsey: *Mont. Not. Roy. Astron. Soc. Geophys. Suppl.*, **5**, 409 (1949).

[62] L. Thomsen: *Journ. Geophys. Res.*, **76**, 1342 (1971).

[63] L. Johnson: *Journ. Geophys. Res.*, **72**, 6309 (1967).

# An Aversion to Inversion.

L. Knopoff

*University of California - Los Angeles, Cal.*

## 1. – Introduction.

Suppose that we have observations of certain properties of the Earth, such as travel-times of body waves, phase velocities of surface waves or periods of free oscillation. We construct a model to try to fit these observations. Let the observations be $f_\alpha$ and let the model parameters be $l_n$. In general, if we try to generate the theoretical values of the quantities $f$ as a consequence of the postulated model $l_n$, we find they do not match the observations. Let the differences between the observed and the computed quantities be called « data » and be given the symbols $g_\alpha$.

Since our fit is not exact, we must amend the model. Let the changes in the model parameters be $\Delta l_n$. Then, by a simple Taylor expansion, we see

$$g_\alpha = \frac{\partial f_\alpha}{\partial l_n} \Delta l_n + \frac{1}{2} \frac{\partial^2 f_\alpha}{\partial l_n \partial l_m} \Delta l_n \Delta l_m + \dots \, .$$

We restrict the problem of determining the $\Delta l_n$ to the solution of the equation obtained by neglecting terms higher in order than the first and obtain

(1)
$$g_\alpha = G_{\alpha n} m_n \, ,$$

where we have let $m_n = \Delta l_n$, $G_{\alpha n} = \partial f_\alpha / \partial l_n$.

We can now state our problem in the following form. Let $g_\alpha$ represent « data » such as travel-times of body waves, periods of free oscillation, phase velocities of surface waves, mass and moment of inertia, etc. These are the result of observations on the surface of the Earth and are suitably represented in discrete form. The data are statistically independent observations. Let $m_n$ be the parameters of the « real Earth » that we wish to find, again suitably represented in discrete form. We let $\mu_n$ be some estimate of $m_n$ that we shall obtain. Further, there exists some theoretical operator $G$ which generates the data from the model. Let this operator be linear in the sense that it does not change significantly as we move from model to model in our search for the properties of a solution. We solve the equation

$$(1) \qquad\qquad g_\alpha = G_{\alpha n} m_n,$$

where these are simultaneous linear equations in the unknowns $m_n$.

In this discussion, Greek indices represent the observations and Latin indices represent the model parameters. The summation convention applies except where explicitly indicated. Let the number of degrees of freedom of the observations be $\Lambda$. Let the number of degrees of freedom of the model parameters be $N$,

$$\alpha, \beta, \gamma, ..., \text{etc} = 1, 2, ..., \Lambda, \qquad l, j, m, n, ..., \text{etc} = 1, 2, ..., N.$$

We discuss several cases including those for which $N$ is less than, equal to or greater than $\Lambda$.

## 2. – $N < \Lambda$.

In this case, we assume that the data are not noise-free. In general, the theoretical values derived from the model will not be consistent with all $\Lambda$ values of the data since the calculated values of the data will have too few degrees of freedom. Thus, we assume that the observations of the $g$'s differ from the theoretical values because of noise, *i.e.*, statistical uncertainty in the observations. We must make some postulate, according to a theory of the errors in the observations, of the residuals in the observations. Often, lacking a more sophisticated theory, the least-squares procedure is invoked; this method leads to simple mathematics. In the least-squares procedure, we assume *a priori* that the residuals $(G_{\alpha n} m_n - g_\alpha)$ are normally distributed. We construct

$$(2) \qquad\qquad \min_m (G_{\alpha n} m_n - g_\alpha)(G_{\alpha k} m_k - g_\alpha)$$

and get

$$G_{\alpha k}(G_{\alpha n} m_n - g_\alpha) = 0.$$

Thus, the solution to (2) is

$$(3) \qquad\qquad m_n = g_\alpha G_{\alpha k}(G_{\beta k} G_{\beta n})^{-1},$$

if the expression in parentheses has an inverse. If weights are introduced into (2), we get solutions other than (3).

It should be recognized that we have constrained the solution by specifying (in depths) where the $N$ unknowns are located. For example, the $g_\alpha$ may be short-period surface wave data, but we may have arbitrarily chosen a *mathematically valid* set of unknowns $m_n$ such as variables representing the properties of the earth's core. Variations in these variables do not have a large influence on the values of $g_\alpha$. For the case $N < \Lambda$, one can always get a solution from (3) but it may be physically unreasonable (*e.g.*, densities of $10^3$ cgs or negative moduli, etc.), since no constraints are placed upon the model $m_n$. These constraints are called by Keilis-Borok « inequalities of state ». The imposition of inequalities of state turns the linear problem (2) into a nonlinear one.

On the other hand, we might have taken the $m_n$ as representing near-surface parameters that would be more appropriate, namely to choose $m_n$ in those places where $G_{\alpha n}$ are large, but no definite rule can be written about this since any assumption about the location of the parameters $m_n$ will lead to a solution, no matter how implausible.

## 3. – $N = \Lambda$.

In this case, we have as many unknowns $m_n$ as we have data. We can take the data to be noise-free.

$$m_n = g_\alpha G_{\alpha n}^{-1},$$

if $G$ has an inverse.

Both cases 2 and 3 are complicated for similar reasons, as cited above. To repeat, we get a different solution for different choices of the places in the earth where the unknown $m_n$ are to be located; no rule can be stated in an unbiased way about determining where these sites are to be chosen, although we may wish to place them where the $G_{\alpha n}$ are large.

## 4. – $\Lambda < N$.

This is a case of major geophysical interest. Our demands for information concerning the earth's interior will always exceed in detail the availability of the quantity of independent data to determine the structure. We assume, in this Section, that the data are noise-free; that is, that repetitions of the geophysical measurements will lead to the same $g_\alpha$.

We note the following problem. If we guess perfectly, *i.e.*, if $l_n$ is the real Earth, then $g = 0$ and $m = 0$, which is consistent with (1). But, in the cases $N > \Lambda$, there are many models which satisfy the data equation. If we start with any one of these models, we also get $g = 0$. But, in this case, we cannot assume that $l_n$ is the real Earth. Thus, some properties of the matrix $G$ must be singular. The properties of the matrix $G$ are the subject of this Section.

If $g \neq 0$, our problem is evidently to take the rectangular matrix $G_{\alpha n}$ and increase its dimensions from $\varLambda \times N$ to $N \times N$ by «inventing» some new data. Since we can always invent new numbers easily, *e.g.*, by some Monte Carlo procedure, the problem is evidently to invent the new data artfully and in a controlled way, so that we know the limitations on the process. When this is done, we will understand the limitations on the validity of the solution.

It is evident by inspection that any choice of fictitious «data» which will permit one to expand and «square up» the matrix $G$ will lead to a solution by the methods of Sect. **3**. Thus, there exists a continuum of solutions to the data equation and it is assumed that the real Earth is one of these solutions.

Consider the space of Earth models $\{M\}$. The solutions to eq. (1) are represented by a hyperspace in $\{M\}$ which is called $\{N\}$ (see Fig. 1). We select an initial model $l_n$ in $\{M\}$ and, by «inventing» some data arbitrarily, we proceed to a solution in $\{N\}$ which is $l_n + \mu_n$. The real Earth also lies on $\{N\}$ in this noise-free case. Different methods of inventing «new data» generate different members of $\{N\}$.

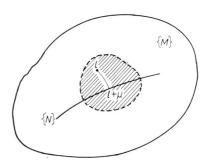

Fig. 1. – Space of Earth models $\{M\}$ and of solutions to the data equation $\{N\}$. The shaded region represents the subspace in which $G$ and hence $\varLambda$ does not change significantly from model to model. If we invent new data inappropriately, we may generate models which lie outside the range of slowly varying $G$.

The following procedure illustrates one method of generating models. Indeed, it generates some of the possible $\mu_n$ in one step from $l_n$ onto $\{N\}$, but not all of the models $\mu$ that can be generated in a one-step process. This special case is

$$(4) \qquad \mu_n = g_\alpha f_{nm} G_{\beta m} (f_{pq} G_{\alpha p} G_{\beta q})^{-1} .$$

That this satisfies

$$(5) \qquad g_\alpha = G_{\alpha n} \mu_n ,$$

follows by inspection. In (4), $f_{nm}$ is an arbitrary symmetric matrix such that the matrix in parentheses is nonsingular. The special case $f_{nm} = \delta_{nm}$ was considered by SMITH and FRANKLIN [1].

It is possible to show that the result (4) is the consequence of finding a solution to the problem

$$(6) \qquad \min f_{kl}^{-1} (m_k - \mu_k)(m_l - \mu_l) ,$$

subject to the condition (5). The proof follows by substituting $\mu_n = b_\alpha G_{\alpha m} f_{mn}$ and minimizing (6) with respect to a choice of the vector $b_\alpha$. The physical significance of the criterion (6) is that we have constructed in (6) a weighted autocorrelation function. Minimizing this is equivalent to minimizing the weighted power spectral differences between the model and the real Earth. Thus, these models have a maximum smoothness in the sense that jumps in differences $(m_n - \mu_n)$ would have a broad spectrum and this would be minimized. In this postulate, we have assumed that the models generated to fit (1) have some stochastic property that permits the application of this smoothing criterion.

The reader may be amused to consider the similar problem

$$\min f_{kl}^{-1}(m_k - f_{km}\mu_m)(m_l - f_{ln}\mu_n) ,$$

where we let $\mu_m = b_\alpha G_{\alpha m}$ and the minimization is taken with respect to a choice of the vector $b_\alpha$. The solution to this problem is

$$\mu_p = g_\beta G_{\alpha p}(f_{km} G_{\alpha m} G_{\beta k})^{-1}$$

and does not satisfy the data equation (5).

The concept of the averaging kernel is useful. This kernel has the property that it generates a model $\mu_n$ as a local linear weighted average of the property $m_n$, namely

(7)                                        $\mu_n = A_{nk} m_k .$

For the models in (4),

(8)                                        $A_{nk} = f_{nm} G_{\alpha m} G_{\beta k}(f_{ij} G_{\alpha j} G_{\beta i})^{-1} .$

This can be demonstrated by substituting (8) into (7) and applying (1), thereby generating (4).

## 5. – Deltaness.

BACKUS and GILBERT [2, 3] proposed that the averaging kernel $A_{ij}$, defined in (7), is an important way to study the relationship among the models in $\{N\}$, i.e., the solutions to (5), including the « real Earth » model. It would be pleasant if $A_{nk} = \delta_{nk}$, for then $\mu_n = m_n$. But $A_{nk}$ is never the Kronecker delta, since $N > \Lambda$, i.e., because of dimensionality arguments. Further

(9)                                        $\det A_{nk} = 0 ,$

a condition, which cannot be satisfied by the Kronecker delta. Equation (9) holds since, if (9) where not true, it would be possible to invert (7) and solve for $m_k$.

BACKUS and GILBERT propose to make $A_{ij}$ as deltalike as possible according to the following prescription. We start at the same point as before. We seek to find some geophysical property such that (1) is satisfied, i.e.

$$(1) \qquad\qquad g_\alpha = G_{\alpha k} m_k \, ,$$

where $G_{\alpha k}$ is a theoretical operator and $g_\alpha$ are some observations. Let $A_{ij}$ be the averaging kernel which relates the desired property $m_k$ to $\mu_k$. The model that will actually be obtained is

$$(7) \qquad\qquad \mu_i = A_{ij} m_j \, .$$

We arbitrarily introduce a new matrix $a_{\alpha n}$ such that

$$(8) \qquad\qquad A_{nm} = a_{\alpha n} G_{\alpha m} \, .$$

By direct substitution of (10) and (1) into (7), we have

$$(11) \qquad\qquad \mu_i = a_{\alpha i} G_{\alpha j} m_n = a_{\alpha i} g_\alpha \, .$$

I shall comment below that the actual model to be obtained by this procedure is not particularly relevant.

For eqs. (12) through (18), we drop the summation convention and sum only where indicated. We normalize each row of the matrix $A$

$$(12) \qquad\qquad \sum_{j=1}^{N} A_{ij} = 1 \, .$$

This condition leads to models $\mu$ which are appropriate for the case $m = \text{const}$ from (7).

We now try to make $A_{ij}$ as close to a Kronecker delta as possible

$$A_{ij} \approx \delta_{ij} \, ,$$

noting, as before, that this cannot be done exactly. To find $a_{\alpha n}$, the generator of $A_{mn}$, we invent a new matrix $J_{ij}$ with certain properties. We minimize the quantity

$$(13) \qquad\qquad \sum_{j=1}^{N} J_{ij} A_{ij}^2 \, ,$$

subject to the constraint (12). The quantities $J$ are arbitrary, but clearly they play the role of weighting factors. It is relatively simple to show that the $i$-th row can be made most deltalike if $J_{ij}$ is chosen so that it is least at the diagonal elements $i = j$ and greatest far from the diagonal, and increases monotonically as one goes away from the diagonal. BACKUS and GILBERT suggest

$$(14) \qquad\qquad J_{ij} = (i-j)^2 \, .$$

Thus, the prescription involves making each row separately as deltalike as possible.

We introduce Lagrangian multipliers $2\lambda_i$ and minimize, from (13) and (12)

$$(15) \qquad\qquad \psi_i = \sum_{j=1}^{N} J_{ij} A_{ij}^2 + 2\lambda_i \sum_{j=1}^{N} A_{ij} \, .$$

From (10), we compute the minimum of

$$(16) \qquad\qquad \psi_i = \sum_{\alpha}^{A} \sum_{j}^{N} J_{ij} (a_{\alpha i} G_{\alpha j})^2 + 2\lambda_i \sum_{\alpha}^{A} \sum_{j}^{N} a_{\alpha i} G_{\alpha j} \, .$$

We differentiate (16) with respect to $a_{\alpha i}$ and thus we solve

$$(17) \qquad\qquad \sum_{\alpha}^{A} \sum_{j}^{N} J_{ij} (a_{\alpha i} G_{\alpha j}) G_{\beta j} + \sum_{j}^{N} \lambda_i G_{\beta j} = 0$$

together with the constraint condition of normalization (12),

$$(18) \qquad\qquad \sum_{\alpha}^{A} \sum_{j}^{N} a_{\alpha i} G_{\alpha j} = 1 \, .$$

Equations (17) and (18) are $(A+1)$ simultaneous equations in the unknowns $a_{\alpha i}$ ($i$ is fixed) and $\lambda_i$ ($i$ is fixed). Solving these simultaneous equations for $a_{\alpha i}$, we repeat the process for each row by changing the index $i$. From the complete matrix $a_{\alpha i}$, we can substitute in (10) and generate $A_{nm}$. If we wish, we can substitute in (11) and generate $\mu_i$, the model, although, as will be shown below, this is not crucial to the discussion.

Although each row of the matrix $A_{nm}$ is made as deltalike as possible, the deltaness of any given row does not depend on the properties of any other row explicitly. Thus, the matrix $A_{nm}$ generated by this procedure is in general nonsymmetric.

We can further ask, does the model $\mu_{ij}$ generated in (11)-(18) satisfy the data equations in a way that the models generated in Sect. **4** did? *If the model* does satisfy (5), namely

(5) $$g_\alpha = G_{\alpha k} \mu_k ,$$

then from (11)

(19) $$g_\alpha = G_{\alpha k} a_{\beta k} g_\beta$$

will be an identity. This will require that

(20) $$G_{\alpha k} a_{\beta k} = \delta_\alpha .$$

Because of the noninteractive properties of the rows of $a_{\alpha i}$, there is no reason for (20) to be satisfied. Hence, the $A_{ij}$ matrix obtained for the deltaness criterion does not generate values $\mu_i$ that satisfy the data equation (5) in general. Further, the $A$-matrix (10) for the deltaness problem is not found among the members of the set (8).

The result that the model $\mu_k$ from (11) does not satisfy (5) is not distressing. I shall point out below that the property of interest is $A_{mn}$ and not $\mu_n$, despite our natural fascination to learn more about the interior of the Earth. If we desire to find a value of $\mu$ by this procedure, we can, in principle, set up the problem as an iteration. We construct the differences

$$g'_\alpha = g_\alpha - G_{\alpha k} \mu_k$$

as a new set of data for a new problem similar to (1)

$$g'_\alpha = G_{\alpha k}(m_k - \mu_k)$$

and hope that the procedure converges. There is no guarantee of convergence of the iteration. Indeed, in many practical cases the process diverges, *i.e.*, the succession of models moves away from a solution that satisfies (5) instead of toward it. The matrix $A_{ij}$ for each of the succession of models in the iteration is the same as the single-stage averaging kernel (10) by virtue of the fact that the matrix $a_{ij}$ does not depend on the data; by (10), it follows that $A_{ij}$ does not depend on the data $g_\alpha$. Thus, we conclude that there is an infinite set of models $\mu$ related to $m$ through (7) with the same $A$; most of these models do not satisfy the data equation. I shall point out that all models $\mu$ which satisfy the data equation (5) are related through the properties of a particular averaging kernel $A$, whether most deltalike or not, for the linear problem. This is independent of whether the model and the averaging kernel were derived by the same method or not.

If we can generate one averaging kernel $A$ by the method of this Section, can we generate others? This is certainly possible in some sense, by changing the weighting function $J_{ij}$. Care must be exercised in changing $J_{ij}$ too strongly; a function such as $(i-j)^{100}$ may introduce instabilities in the simultaneous solution of (17) and (19).

## 6. – Other solutions: $\Lambda < N$.

At the outset of Sect. **4**, it was stated that it was easy to find solutions $\mu_n$ to (5) merely by « inventing » data. We are evidently surfeited with solutions. In the solutions of eq. (4), a family of such solutions has been generated through the choice of the weighting functions $f_{ij}$. These solutions have the property of providing maximum smoothing in a spectral sense. The problem may also be solved through eqs. (17) and (18) and an iteration without guarantee of convergence, but the averaging kernel will have maximum deltaness.

Perhaps the simplest of solutions, if solutions are all that are sought, is to interpolate among the data to obtain $N-\Lambda$ new « data » points and to solve the problem by the method of Sect. **3**.

The iteration of the deltaness problem can be circumvented and a modification of the deltaness method can be obtained which proceeds to a solution $\mu$ on $\{N\}$ in one step. Let us append to the problem (16) the additional constraint that the data equations (5) are to be satisfied. This constraint is written as a condition on the matrix elements $a$ in (20), since (20) now holds. In (20), the the matrix $a$ is a dual of $G$; $a$ is not $G^{-1}$ because $\Lambda \neq N$. The contraints are explicitly stated as (19).

We write, using our summation signs explicitly

$$\min \sum_{\alpha}^{\Lambda} \sum_{j}^{N} J_{in}(a_{\alpha i} G_{\alpha j})^2 + 2\lambda_i \sum_{\alpha}^{\Lambda} \sum_{j}^{N} a_{\alpha i} G_{\alpha j} + 2 \sum_{\beta=1}^{\Lambda} \sum_{j=1}^{N} \sum_{\alpha=1}^{\Lambda} \pi_\beta g_\alpha a_{\alpha j} G_{\beta j}$$

for our minimization problem, with $\lambda$ and $\pi$ the Lagrangian multipliers. Differentiating with respect to $a_{\alpha i}$, we get

$$(21) \qquad \sum_{\beta}^{\Lambda} \sum_{j}^{N} J_{\alpha j} a_{\beta i} G_{\alpha j} G_{\beta j} + \lambda_i \sum_{j}^{N} G_{\alpha j} + \sum_{\beta}^{\Lambda} \pi_\beta g_\alpha G_{\beta i} = 0 .$$

These are now $N\Lambda$ simultaneous equations in the $N$ unknowns $a_{\beta i}$, which together with $\Lambda$ constraints of type (19) and the $N$ constraints of type (18) form a complete set, since there are $N$ multipliers $\lambda$ and $\Lambda$ multipliers $\pi$. We solve (21), (19) and (18) for the $a_{ij}$. The solution to this problem, that of maximum deltaness with the constraint that (5) be satisfied, cannot lead to an averaging kernel that is as deltalike as that without the latter constraint. It is clear, in view of (20) and (10), that $\Lambda$ is the product of $G$ with its dual $a$, the dual being defined in (20).

## 7. – Which solution is « best »?

Hints have been given above in the discussion of the deltaness criterion of eqs. (16)-(18) that perhaps the obtaining of a solution $\mu_n$ to the data equation

$$(5) \qquad\qquad g_\alpha = G_{\alpha n}\mu_n$$

may not be of the greatest importance. Let us investigate why.

Glutted with solutions, we approach the problem established at the outset of this Section, namely, which one is the « best »? The response to this problem must be whichever solution has the most aesthetic appeal. The question is meaningless since, without additional insight, there is no way to select from among the solutions the appropriate one, $m_n$.

However, we inquire what properties the solutions have in common. From (1) and (5), we see

$$(22) \qquad\qquad G_{\alpha n}\mu_n = G_{\alpha n}m_n .$$

Let us take the scalar product of (22) with the matrix $a_{\alpha m}$ defined in (10) or with $f_{mp}G_{\beta p}(f_{ij}G_{\beta j}G_{\alpha i})^{-1}$ as in (8). From (8) and (10), we have

$$(23) \qquad\qquad A_{mn}\mu_n = A_{mn}m_n .$$

Thus, the weighted average $A_{mn}\mu_n$ of any solution is the same as that for any other solution, including the real Earth. Hence, the averaging kernels with maximum deltaness, corresponding to the problem (12)-(18), give the best resolution possible for the determination of the properties of the interior where these properties are now expressed as weighted averages. Where $A_{mn}$ is most like $\delta_{mn}$, all solutions, including those generated by the methods outlined above, will have properties close to one another and close to that for the Earth. If $A_{mn}$ in its optimum circumstance is not very deltalike, then this remains the best we can do: namely, the local average of all solutions $\mu_n$ will be an average over a broad range of indices $m$; we cannot hope to learn more, for the given data set $g_\alpha$.

Thus, one need not be concerned about the failure of the deltaness procedure to provide a solution. Any solution that satisfies (5) will have the same local average properties as the real Earth, no matter how the solution was derived. There are many averaging kernels $A_{mn}$, as for example those given by (8). These averaging kernels provide correct local averages for all solutions, whether the solution was derived by the method which yielded the corresponding $A_{mn}$ or not. The most deltalike kernels $A_{mn}$ of (17)-(18) provide the « best » local average estimates, i.e., involve averaging over as small a range of model parameters as possible and apply to any solution. Thus, the best that can be said for any of the solutions is that they can be obtained by the explicit methods listed above; what is more important is the local average, however.

The linear nature of the problems referred to above implies that the partial derivatives $G_{\alpha n}$ do not change significantly over the range of models, and hence that all models will have the same $A_{mn}$. Suppose that we reduce the nonlinear problem to a linear one by the severe application of constraints such as « inequalities of state ». Then the local deltaness of some $A_{mn}$ will not necessarily mean that we have a good estimate of the particular average property for the real Earth.

## 8. – $\Lambda < N$: the simultaneously underdetermined and overdetermined case.

Suppose that $\Lambda < N$ as before, but that the data are not noise-free. Then, from eq. (5),

$$\delta g_\alpha = G_{\alpha n}(\delta \mu_n) \,,$$

where $\delta g_\alpha$ and $\delta \mu_n$ are the uncertainties in the data and the corresponding ones in the model. KEILIS-BOROK and KNOPOFF [5] have approached this problem by attempting to find the class of models $\mu_n \pm \delta \mu_n$ which satisfies this equation by a search through an $N$-dimensional parameter space, whether the problem is linear or not. BACKUS and GILBERT [2] propose that the problem be studied by computing the debilitation of the deltaness, i.e., the broadening of the $A_{nm}$ matrices, and hence the loss of resolution in the determination of the average properties of the Earth structure.

We estimate the variance in the model

(24),                          $\varepsilon_{ii} = \delta \mu_i \delta \mu_i$        (summation convention does not apply) .

From (11), this quantity is

(25)                          $\sum_{\alpha,\beta} a_{\alpha i} \delta g_\alpha a_{\beta i} \delta g_\beta = \varepsilon_{ii} \,.$

It is evidently not possible to minimize both (24) and (16) at the same time, but it is possible to minimize a linear combination of these two expressions, namely

(26)        $\min \left\{ \sum_{\alpha}^{\Lambda} \sum_{j}^{N} J_{ij}(a_{\alpha i} G_{\alpha j})^2 + 2\lambda_i \sum_{\alpha}^{\Lambda} \sum_{j}^{N} a_{\alpha i} G_{\alpha j} + f \sum_{\alpha,\beta}^{\Lambda} a_{\alpha i} a_{\beta i} \delta g_\alpha \delta g_\beta \right\} \,,$

with respect to a choice of the matrix $a_{\alpha i}$ for a given weighting factor $f$.

It should be noted that the quantity being minimized in (25) is not the same as that minimized in (6) with the factor

$$f_{kl} = \delta_{kl} \, ,$$

since in (6) we have summed over the repeated indices. BERRY and KNOPOFF [4] have given one example of the calculation of $\varepsilon_{ii}$ from surface wave data.

## REFERENCES

[1] M. L. SMITH and J. N. FRANKLIN: *Journ. Geophys. Res.*, **74**, 2783 (1969).
[2] G. BACKUS and J. F. GILBERT: *Geophys. Journ. Roy. Astron. Soc.*, **16**, 169 (1968).
[3] G. BACKUS and J. F. GILBERT: *Phil. Trans. Roy. Soc.*, **266**, 123 (1970).
[4] M. J. BERRY and L. KNOPOFF: *Journ. Geophys. Res.*, **72**, 3613 (1967).
[5] V. I. KEILIS-BOROK and L. KNOPOFF: (1971), in preparation.

# Attenuation.

*University of California - Los Angeles, Cal.*

## 1. – Introduction.

Observations of seismic wave attenuation in the Earth are not easily obtained. We would like to know the ability of a single volume element of the interior of the Earth to extract energy from elastic waves passing through it. But, of course, we cannot extract a single volume element from the deep interior of the Earth and subject it to laboratory experiments. We can perform laboratory experiments of attenuation on materials which are possible constituents of the interior, but these experiments cannot be done at the high pressures and temperatures which are found through most of the Earth's interior. What we would like to do with attenuation measurements for the Earth is to use them as diagnostics for compositional and state studies, *i.e.*, as input to the generalized inverse problem.

Because we cannot study a single volume element, our measurements of attenuation at the surface of the Earth must perforce be operations which integrate over a large region of the Earth. Thus, much of the discussion here will be concerned with deconvoluting a complex operator.

Before embarking upon our task, let us consider the units of attenuation. KNOPOFF and MACDONALD [1] introduced the electrical engineer's symbol $Q$ into the geophysical literature which is called the specific attenuation factor. This is a convenient symbol because it is dimensionless. A measure of $Q$ is given by the relation

$$Q^{-1} = \frac{\Delta E}{2\pi E},$$

where $\Delta E$ is the amount of elastic energy dissipated per cycle of harmonic excitation in a certain volume and $E$ is the peak elastic energy of the system in the same volume; often the potential energy locally is not equal to the kinetic energy because of dispersion and so this may be a difficult version to use.

In addition, definitions of $Q$ can be written for damped harmonic space or time-rate processes such that the logarithmic decrement is $\pi/Q$. Consider damped propagating waves

$$\exp\left[-\alpha x\right]\exp\left[i\omega(t-x/c)\right],$$

with frequency $\omega$, phase velocity $c$ and attenuation factor $\alpha$. These quantities are related to the decrement by

$$Q_x^{-1} = 2\alpha c/\omega .$$

For standing waves at a fixed point in space, the wave function is $\exp\left[-\gamma t\right]\cdot$ $\cdot\sin\left(\omega t+\delta\right)$. The relation between the standing wave decrement $\pi/Q_T$ and the decay rate $\gamma$ is

$$Q_T^{-1} = \frac{2\gamma}{\omega} .$$

The two specific attenuation factors $Q_T$ and $Q_x$ are not equal to one another for dispersive waves in the presence of weak attenuation. It was shown by BRUNE [2] and KNOPOFF et al. [3] that $Q_x$, the $Q$ measured from propagating waves, and $Q_T$, the $Q$ measured from standing waves, are related by

$$cQ_x = UQ_T ,$$

with $U$ the group velocity. The proof is elementary.

Because they are the invariants of the strain tensor, it is useful to consider deformations in bulk and in shear for isotropic materials as fundamental for both elastic and anelastic parameters. We consider the bulk modulus $K$ and the bulk specific attenuation factor $Q_K$ and the corresponding shear properties $\mu$ and $Q_s$ in relation to the properties to be expected for $P$-waves. Since there are only two independent elastic and anelastic parameters for isotropic materials, all others are derivable from these. Seismic body waves do not excite materials in bulk deformation modes; instead, the irrotational or $P$-waves are associated with a different modulus $(\lambda+2\mu)$; we consider the relation between $Q_p$, the specific attenuation factor for $P$-waves, and $Q_K$ and $Q_s$.

The wave equation for $P$-waves in a homogeneous isotropic material is

$$(\lambda+2\mu)\nabla^2\varphi + \omega^2\varrho\varphi = 0 ,$$

with $\varrho$ the density and $\varphi$ the time Fourier transform of the wave function. The wave velocity is

$$V_p = \left(\frac{\lambda+2\mu}{\varrho}\right)^{\frac{1}{2}} .$$

Attenuation can be introduced into the system by making $(\lambda + 2\mu)$ or $\varrho$ or both complex. Since we associate attenuation with imperfect elasticity, we choose the first of these and write

$$(\lambda + 2\mu)_0 \left(1 \pm \frac{i}{Q_p}\right) \nabla^2\varphi + \omega^2\varrho\varphi = 0 ,$$

since this, as is easily shown, gives the expression

$$\alpha = Q_p^{-1}\omega/2V_p$$

in correspondence to the expression above, if $Q_p \gg 1$; the sign is chosen appropriately to give attenuation. Thus, if we write the complex elastic modulus as the perturbation

$$\lambda + 2\mu = (\lambda + 2\mu)_0 + i\delta(\lambda + 2\mu) ,$$

we see that

$$\frac{1}{Q_p} = i\,\frac{\delta(\lambda + 2\mu)}{(\lambda + 2\mu)_0} .$$

Thus, we can relate the $Q$'s in $P$-waves, in $S$-waves and in bulk processes through

$$\frac{1}{Q_p} = i\,\frac{\delta(\lambda + 2\mu)}{(\lambda + 2\mu)_0} , \qquad \frac{1}{Q_s} = i\,\frac{\delta\mu}{\mu_0} , \qquad \frac{1}{Q_K} = i\,\frac{\delta K}{K_0} .$$

But $K = \lambda + \frac{2}{3}\mu$ for isotropic materials. Since

$$\frac{\delta(\lambda + \frac{2}{3}\mu)}{\lambda + \frac{2}{3}\mu} = \frac{\delta(\lambda + 2\mu - \frac{4}{3}\mu)}{\lambda + 2\mu - \frac{4}{3}\mu} ,$$

then

$$\frac{1}{Q_K} = \frac{1}{Q_p}\,\frac{3(1-\sigma)}{1+\sigma} - \frac{1}{Q_s}\,\frac{2(1-2\sigma)}{1+\sigma} ,$$

where $\sigma$ is Poisson's ratio. For the Poisson condition, $\sigma = \frac{1}{4}$, and

$$\frac{1}{Q_K} = \frac{9}{5Q_p} - \frac{4}{5Q_s}$$

as has been shown by KNOPOFF [4].

   Creep in solids is probably related to attenuation in solids by virtue of the fact that the principal agents in both processes are associated with deformation in the presence of defects in crystal structure, such as impurities or crystal dislocations; the scale of the deformations is much larger in creep processes.

permanent creep is almost never observed in deformation experiments in which the sample is compressed hydrostatically but not fractured. It is thus plausible to take $Q_K^{-1} = 0$. The relation $Q_p \approx \frac{9}{4} Q_s$ is probably a good approximation.

## 2. – Observations on the Earth.

The basic measurements on attenuation that can be made on the Earth include those on body waves, surface waves and on the free vibrations. The simplest type of body wave experiment is to measure the quantity

$$\exp\left[-\int \frac{\pi\,ds}{TQ_{p,s}(\alpha,\beta)}\right],$$

for either $P$- or $S$-waves along the path of a ray, after effects of transmission or reflection from boundaries have been taken into account.

The simplest experiment is to measure body waves reflected between two perfect reflectors. As the simplest case, we can consider normal incidence of $S$-waves between the core and the surface of the Earth. If the wavelength is short compared with the distance over which changes in velocity and density take place, then the comparison of amplitudes of multiply-reflected waves gives a measure of the attenuation, integrated over the radius of the Earth, as above. The sharp boundary at the Moho, and elsewhere, can be taken into account by computing a theoretical reflection coefficient. For a particular deep focus earthquake in South America, the comparison of such multiple reflections for waves traveling both upward and downward from the shock led ANDERSON and KOVACH [5] and STEINHART et al. [6] to conclude that the upper mantle was a significantly stronger attenuator than was the lower mantle. For the upper mantle $Q_s \simeq 150$, while for the lower it is $Q_s \simeq 1500$.

Such measurements for $P$-waves are much more difficult to perform since the core-mantle boundary is not a perfect reflector. However, $Q_p$, in principle, can be derived from the expression $Q_p \approx \frac{9}{4} Q_s$ above. Regrettable as this failure to measure $Q_p$ accurately may be,—many almost reliable attempts have been made—, $Q_s$ information is extremely useful since $S$-waves are much stronger indicators of the thermal state of the mantle than are $P$-waves.

Surface wave attenuation has been measured by BEN-MENAHEM [7] for both Love and Rayleigh waves for globe-circling paths. The « Love » waves show a gentle decrease in $Q^{-1}$ toward the shorter periods; the Rayleigh wave $Q^{-1}$ increases from 300 s to 100 s and then decreases gently at shorter periods (Fig. 1). $Q$'s from line widths of the free modes of oscillation have been reported by SMITH [8] and NESS et al. [9]. Other measurements are not as reliable. A summary of experimental results is given by KNOPOFF [4].

The interpretation for Love waves proceeds as follows: Construct the dispersion equation for the wave number

$$k = k[\omega, \mu(r), \varrho(r)] \, .$$

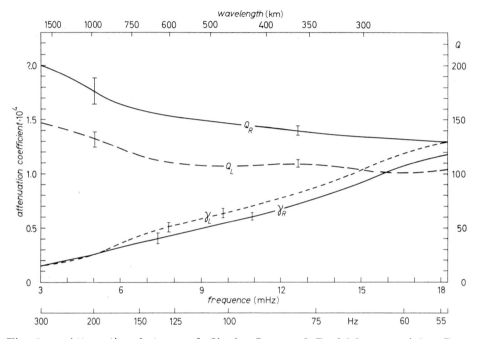

Fig. 1. – Attenuation factors and $Q$'s for Love and Rayleigh waves (after BEN-MENAHEM [7]).

Now perturb this expression by varying $\mu(r)$. We get, at fixed frequency,

$$\mathrm{Im} \, \delta k = \int \mu \, \frac{\partial k}{\partial \mu} \, \mathrm{Im} \, \frac{1}{\mu} \, \frac{\partial \mu}{\partial r} \, \mathrm{d}r \, .$$

Thus, since the apparent $Q$ is $\frac{1}{2} (\mathrm{Im} \, k/k)$, we have

$$\frac{1}{Q_x(\omega)} = \int\limits_{core}^{surface} \frac{\mu}{k(\omega)} \, \frac{1}{Q_s(\omega)} \, \mathrm{d}\left(\frac{\partial k(\omega)}{\partial \mu}\right) \, .$$

This expression is, unfortunately, a Fredholm equation of the first kind and has no unique solution, as we have seen in the discussion of inversion. The best we can do is find out something about the averaging kernels and their

properties. Further, we must make some assumption concerning the dependence of the intrinsic $Q$ upon $\omega$. If we make the assumption (see below) that $Q$
is independent of $\omega$, then, from the Love-wave data, we get results fully consistent with the $S$-wave results described above (KNOPOFF [4]), but of course
not uniquely so. The data are not strongly discriminatory as to $Q_s(r)$.

The inability of $Lg$-waves to be transmitted across even as small an ocean
as the Mediterranean shows $Q_s$ in the low-velocity channel to be $Q_s < 100$
(about) for oceans (SCHWAB and KNOPOFF, [10]).

It seems plausible to associate the low-velocity channel of the earth with
a low $Q_s$, or a high attenuation in shear. Thus, we might assume that the lithosphere above the low-velocity channel, as well as the material below the channel, has a relatively low attenuation or high $Q$.

Why should this correlation arise? We might expect that partial melting
in the mantle has associated with its attenuation in the fluid parts. A recent
calculation by GARBIN [11] shows significant attenuation, even for small
amounts of material molten along grain boundaries. Thus, we might expect
a larger attenuation where partial melting is more fully developed. In the
western United States, the $S$-wave velocity in the channel is about 4.25 km/s,
while in the eastern United States, it is about 4.45 km/s. Thus, we might expect to see a higher attenuation of seismic waves in the western U.S. than
in the shield areas of the eastern U.S. The observations of the nuclear explosion GNOME gave a much more significant transmission eastward from New
Mexico than westward for waves that penetrated as deeply as 100 km or more.

## 3. – Observations of attenuation at $P = 0$.

A large amount of experimental evidence has been compiled to date regarding the frequency-dependence of $Q$. At STP, most observations of $Q$ at low
frequencies show $Q$ to be independent of frequency. This result has been found
in the field for waves propagated in homogeneous strata *in situ* and in soils.
In the laboratory, it has been found for rocks, pure minerals, glasses, metals
and nonmetals alike. At high frequencies, Rayleigh scattering becomes important, due to finite grain size, and thus $Q^{-1}$ varies as $f^3$—or the attenuation
factor varies as $f^4$.

If the conclusion for low frequencies and near-surface conditions holds
exactly, it is most remarkable, since, on the basis of any linear theory, $Q^{-1}$
should vary as some odd power of the frequency. Consider the perturbed
equation of one-dimensional attenuated wave motions

$$\frac{\partial^2 \psi}{\partial x^2} - \frac{1}{c^2} \frac{\partial^2 \psi}{\partial t^2} = \sum_{mn} a_{nm} \frac{\partial^{2n+2m+1} \psi}{\partial t^{2n+1} \partial x^{2m}}.$$

KNOPOFF and MACDONALD [1] have shown that any combination of linear perturbation operators of the type shown on the right-hand side must lead to an attenuation factor $\alpha \sim \sum b_p \omega^{2p}$ where $a_{nm}$ is small in some sense. The differentiation on the right-hand side with respect to $t$ is odd to ensure dissipation, *i.e.*, time irreversibility, while the differentiation with respect to $x$ is even for space isotropy or space reversibility. The observation $\alpha \sim \omega^1$ for solids in the laboratory is clearly inconsistent with this result, since the observation holds for a wide range of frequencies and for a large number of materials. For the result $\alpha \sim \omega^1$ to hold exactly for all materials would require that the microscopic mechanisms corresponding to each of the $a_{nm}$ be reproduced in each material in exactly the same ratio.

If we set ourselves the problem of finding the operator $f(\psi, x, t)$ in the equation

$$\frac{\partial^2 \psi}{\partial x^2} - \frac{1}{c^2}\frac{\partial^2 \psi}{\partial t^2} = f(\psi, x, t) ,$$

as a real-space-time representation of the forces producing the attenuation, we are faced with several problems. First, of course, is the fact that the attenuation factor must vary as $\omega^1$ when we take the Fourier transform of this equation. No amount of representations of the Fourier transform of the operator $f$ as an operator in the frequency domain, as some investigators have done, will lead to an understandable real-space-time operator $f$. Second is the apparent incompatibility with the linear operators $f$ already discussed. Third, we must recognize that $f$ is indeed linear in $\psi$. Fourth, the Kramers-Kronig, or Paley-Wiener, or dispersion relations must hold. These state that the wave number $k$ must be complex such that

$$k = k_R + ik_I$$

and that $k_R \to 0$, $k_I \to 0$ as $\omega \to 0$ and $k_R$ is the Hilbert transform of $k_I$. Thus, $\alpha \sim \omega^0$ cannot hold for all frequencies down to $\omega = 0$ if $c = $ constant (FUTTERMAN, [12]) over this same range, $k_R = \omega/c$.

All this suggests that $f$ is an operator which is nonlinear in $x$ and $t$ but linear in $\psi$. In this case, let us look at the attenuation of a harmonic wavetrain in the frequency domain. We will see an attenuation of the main component but a relative increase in the intensity of the spectral components outside this central value of frequency. Thus, the mathematics suggests a physics which involves friction; friction is well known to involve a nonlinear process. Indeed the friction at dislocations which migrate under the influence of the incident shear wave is sufficient to generate the properties we want. As bonds are made and broken, high frequency vibrations are generated that remove energy at these high frequencies from the low-frequency acoustic beam.

A solution to the problem of the operator $f$ can indeed be found. For a macroscopic simple harmonic oscillator, we have

$$\ddot{x} + \omega^2 x = g(x, t) ,$$

with

$$g(x, t) = b|x - x_0| \operatorname{sgn} \dot{x} ,$$

where $x_0$ is the amplitude of the oscillation at the moment of reversal of the sign of $\dot{x}$. Of course, this makes the response complicated for nonharmonic wave forms. In this case, a biharmonic spectral analysis (HASSELMANN $et\,al.$ [13]) is recommended for the verification of the nonlinearity of the process. This has recently been done on Rayleigh wave pulses in granite (GREEN [14]) which had earlier been demonstrated to have the usual properties (KNOPOFF and PORTER [15]), $\alpha \sim \omega^1$ at low frequencies.

## 4. – Attenuation at high $P$ and $T$.

The above discussion would seem to provide a justification for the use of $Q(\omega) \sim \omega^0$ in the inversion formula for surface waves. But dislocation migration cannot be a potent mechanism at high pressures and temperatures, $i,e.$, at great depth in the Earth. The mechanism above was proposed by GRANATO and LÜCKE [16], namely that an applied shear stress can tear dislocations away from defects or impurities where they are usually pinned by an attractive force. However, for high temperatures, $Q$ becomes high because the dislocations and impurities themselves become mobile. At temperatures above about 700 °C in MgO, this effect cannot be too important (SOUTHGATE $et\,al.$ [17]). JACKSON and ANDERSON [18] give a review of mechanisms that are likely to be important at depth in the Earth. Among their candidates are 1) relaxation of stress at grain boundaries and 2) « high-temperature background friction ».

Both of these mechanisms are thermally excited processes; this is evidently important in a thermal environment in which materials are close to the melting point, a condition which is obtained for the mantle.

Stress relaxation processes such as diffusion of defects in crystals are mechanisms which act to relieve stress concentration by internal action within an atomic structure. Thus, the principal macroscopic manifestation of such processes is creep or other long period deformation. This process is dissipative and thus might be expected to cause elastic wave attenuation as well. ZENER [19] assumed that the rate of stress relief is proportional to the stress at constant strain. He found that

$$Q_s^{-1} = \frac{\mu_x - \mu_0}{\mu_x} \frac{\omega/\omega_0 \exp\left[-E/kT\right]}{1 + (\omega/\omega_0 \exp\left[-E/kT\right])^2} ,$$

where $kT$ is the Boltzmann energy, $E$ is called an activation energy and $\omega_0$ is a scaling frequency. Here, $\mu_x$ is an unrelaxed shear modulus and $\mu_0$ is the corresponding « relaxed » value. In metals, grain boundary relaxation occurs with peaks in the effect at temperatures of the order of 300 °K to 1000 °K near 1 Hz. Experiments on some oxides in the laboratory show this to be an important process especially if inpurities are present. Since mantle materials are polycrystalline and well-endowed with impurities, this may be an important process for the Earth. Unfortunately, values of $(\omega_0, E)$ for mantle materials are not at all well-known, and especially their variation with pressure. Increase of pressure should also increase $Q$, since mobility of atoms along grain boundaries should be reduced.

JACKSON and ANDERSON [18] propose that the high-temperature background internal frict on observed by metallurgists also applies to mantle materials. This must be a composite of many processes due to vacancy creation and diffusion. This follows a law

$$Q_s^{-1} = A\omega^{-1} \exp\left[-E/kT\right].$$

This is consistent with the attenuation for a macroscopic body with « Maxwellian » viscoelasticity (KNOPOFF and MacDONALD [1]). $Q^{-1}$ increases rapidly as the melting point is approached. It has been observed in oxides.

The inversion of surface wave attenuations with terms which have these complex frequency dependences is difficult but not wholly impossible by numerical procedures, but with additional uncertainties in the reliability of the results.

All of the observations point to the result that attenuation in the low-velocity channel is likely to be, comparatively, rather large. We should look at processes associated with partial melting much more carefully.

## 5. – Partial melting.

GARBIN [11] has looked at the problem of scattering of long-wavelength elastic waves from a homogeneous medium completely permeated by flaws of a circular shape, randomly oriented and with several boundary conditions along the faces. This may be a first approximation to grain boundary melting processes. His results are summarized as follows:

Let $a$ be the radius of the circular crack. Let $N$ be the number of cracks in a volume $V$. Let the elastic moduli of the unflawed medium be $(\lambda_0, \mu_0)$.

For the boundary conditions of vanishing normal stress on both faces of any crack, the effective moduli are

$$\frac{1}{\lambda + 2\mu} = \frac{1}{\lambda_0 + 2\mu_0} \left\{ 1 + \frac{16}{3} \frac{Na^3}{V} \left[ \frac{16}{15} \frac{\mu_0}{3\lambda_0 + 4\mu_0} + \frac{\frac{1}{3}\lambda_0 + \frac{1}{5}\mu_0}{\lambda_0 + \mu_0} + \frac{1}{4} \frac{\lambda_0^2}{\mu_0(\lambda_0 + \mu_0)} \right] \right\},$$

$$\frac{1}{\mu} = \frac{1}{\mu_0} \left\{ 1 + \frac{32}{15} \frac{Na^3}{V} \frac{\lambda_0 + 2\mu_0}{3\lambda_0 + 4\mu_0} \right\}.$$

For ideal liquid boundary conditions on the crack

$$\frac{1}{\lambda + 2\mu} = \frac{1}{\lambda_0 + 2\mu_0} \left\{ 1 + \frac{256}{45} \frac{Na^3}{V} \frac{\mu_0}{3\lambda_0 + 4\mu_0} \right\},$$

$$\frac{1}{\mu} = \frac{1}{\mu_0} \left\{ 1 + \frac{32}{15} \frac{Na^3}{V} \frac{\lambda_0 + 2\mu_0}{3\lambda_0 + 4\mu_0} \right\}.$$

The attenuation factors can be calculated by comparing the energy scattered to infinity with the energy in the incident beam. For $P$-waves and liquid boundary conditions

$$\frac{2\pi}{Q_p} = \frac{Na^3}{V} \left( \frac{128}{45} \right)^2 (k_s a)^3 \left( \frac{\sigma}{\gamma + 1} \right)^2 \left( \sigma^5 + \frac{3}{2} \right).$$

For liquid boundary conditions

$$\frac{2\pi}{Q_s} = \frac{Na^3}{V} \left( \frac{8}{3} \right)^3 \frac{1}{15} (k_s a)^3 \frac{\sigma^5 + \frac{1}{2}}{(\gamma + 1)^2}.$$

In these expressions $\sigma = V_s/V_p$, $\gamma = 2(1 - \sigma^2)$; while $k_s = \omega/V_s$, $k_p = \omega/V_p$. The dependence upon $\omega^3$ shows the process to be a Rayleigh scattering process.

Viscous boundary conditions may also be introduced. These are « impedance boundary conditions », namely that the shear stress across the faces is a quantity $i\eta\omega/h$ times the difference in displacements across the faces, where $h$ is now the thickness of the crack, $\eta$ is the viscosity of the fluid and $i$ introduces a 90° phase shift. The result for $Q$ in this case shows that $1/Q$ varies as $\omega^1$ for small $\eta/h$ in suitable units, while for large $\eta/h$ it varies as $\omega^{-1}$. Thus, the dominant term for producing attenuation in this case, for viscous boundary conditions, is the viscous drag on the faces of the crack. In the case of cracks filled with an ideal fluid, the principal mechanism for loss is Rayleigh scattering.

# REFERENCES

[1] L. KNOPOFF and G. J. F. MACDONALD: *Rev. Mod. Phys.*, **30**, 1178 (1958).

[2] J. N. BRUNE: *Bull. Seismol. Soc. Amer.*, **52**, 109 (1962).

[3] L. KNOPOFF, K. AKI, C. B. ARCHAMBEAU, A. BEN-MENAHEM and J. A. HUDSON: *Journ. Geophys. Res.*, **69**, 1655 (1964).

[4] L. KNOPOFF: *Rev. Geophys.*, **2**, 625 (1964).

[5] D. L. ANDERSON and R. L. KOVACH: *Proc. Nat. Acad. Sci.*, **51**, 168 (1964).

[6] J. S. STEINHART, T. J. SMITH, I. S. SACHS, R. SUMNER, Z. SUZUKI, A. RODRIGUEZ, C. LOMNITZ, M. A. TUVE and L. T. ALDRICH: *Carnegie Institution of Washington Yearbook*, **62**, 280 (1963).

[7] A. BEN-MENAHEM: *Journ. Geophys. Res.*, **70**, 4641 (1965).

[8] S. W. SMITH: *An investigation of the earth's free oscillations*, Ph.D. Thesis, California Institute of Technology, Pasadena (1968).

[9] N. F. NESS, J. C. HARRISON and L. B. SLICHTER: *Journ. Geophys. Res.*, **66**, 621 (1961).

[10] F. A. SCHWAB and L. KNOPOFF: *Surface waves on multilayered anelastic media*, *Bulletin Seismological Society of America*, **61**, Aug. 1971.

[11] H. D. GARBIN: *The diffraction of elastic waves by circular discs*, Ph.D. thesis, University of California, Los Angeles (1970).

[12] W. I. FUTTERMAN: *Journ. Geophys. Res.*, **67**, 5279 (1962).

[13] K. HASSELMANN, W. H. MUNK and G. J. F. MACDONALD: *The bispectra of ocean waves*, in *American Statistical Society Symposium on Time Series*, edited by M. ROSENBLATT (New York, 1963), p. 125.

[14] J. H. GREEN: *Bispectral analysis of small amplitude stress waves in solids*, M.S. Thesis, University of California, Los Angeles (1968).

[15] L. KNOPOFF and L. D. PORTER: *Journ. Geophys. Res.*, **68**, 6317 (1963).

[16] A. GRANATO and K. LÜCKE: *Journ. Appl. Phys.*, **27**, 583 (1956).

[17] P. D. SOUTHGATE, K. S. MENDELSON and P. C. DE PERRO: *Journ. Appl. Phys.*, **37**, 206 (1966).

[18] D. D. JACKSON and D. L. ANDERSON: *Rev. Geophys.*, **8**, 1 (1970).

[19] C. ZENER: *Elasticity and Anelasticity of Metals* (Chicago, 1948).

# Shock-Wave Equations of State of Minerals (*).

T. J. Ahrens

*California Institute of Technology, Seismological Laboratory - Pasadena, Cal.*

## 1. – Introduction.

Shock-wave data play a unique role in the study of the interior of the Earth in that at pressures of 500 kbar to 3700 kbar, or at equivalent depths in the Earth of 1200 km and 6400 km (center), dynamic techniques provide the only means of studying mineral properties in the laboratory. The relation between shock pressure ($p$), shock-induced density ($\varrho$), and internal energy ($e$), along a curve called the Hugoniot, is the form of an equation of state for solids or fluids which is usually obtained with shock-wave techniques. Shock-wave data in this form have been obtained for mantle and core materials and are useful in a study of the constitution of the Earth's interior in the following two ways:

1) The Hugoniot curves, upon suitable reduction, provide pressure-density-temperature states for materials; these impose important constraints upon Earth models obtained from geophysical and geochemical theories. One form of the equation of state obtained from shock-wave data are the pressure-density isentropes (constants entropy curves). Upon differentiation, the isentropes provide knowledge of the seismic parameter, $\varphi = (\partial p/\partial \varrho)_s$. Since $\varphi$ is obtained directly from seismic determination of the compressional, $v_p$, and shear velocity, $v_s$, at a given depth, or pressure, in the Earth by $\varphi = v_p^2 + \frac{4}{3}v_s^2$, direct comparison between the properties of rocks and minerals in the laboratory under similar pressures and temperatures as are present in the Earth, is possible.

2) The available shock-wave data for silicate minerals indicate that a wide range of pressure-induced phase changes can be delineated. The subse-

(*) Contribution No. 1938, Division of Geological and Planetary Sciences, California Institute of Technology, Pasadena, Cal.

quent measurement of the equation of state of the resulting shock-induced high-pressure phases has provided a useful guide to the probable mineralogy of the lower mantle.

The application of the Rankine-Hugoniot equations, which relate the thermodynamic variables induced by a finite-amplitude stress wave, to the kinematic parameters of the high-speed flow, has long been applied to the study of shock propagation in fluids (see, for example, COURANT and FRIED-RICHS [1]). In an important 1955 paper, WALSH and CRISTIAN [2] demonstrated that the states produced by intense (500 kbar) mechanical shock-waves in metals could be obtained using the Rankine-Hugoniot equations. They showed that the pressure-density relations, obtained by performing a series of experiments on metals, when reduced to isothermal conditions, gave isotherms which agree closely with those which had been obtained statically by BRIDG-MAN up to 100 kbar a few years before. In a latter series of papers by WALSH et al. [3], McQUEEN and MARSH [4], and AL'TSHULER et al. [5], AL'TSHULER et al. [6], shock-wave equation of state data were reported for a host of metallic elements to pressures of, in some cases, approaching 9 megabar (Mbar). A considerable body of data for compounds as well as rocks and minerals, which began to appear in 1955, is collected in the Compendium of Shock-Wave Data (van THIEL et al., [7]). Recently, the dynamic pressure range for which shock-wave data have been obtained has been dramatically extended to 34 Mbar. Shock pressures of this magnitude were reported in the Soviet Union by AL'TSHULER et al. [8], and were obtained by placing appropriate sample assemblies in the vicinity of an underground explosion.

## 2. – Experimental principles.

2'1. *Generating shock-waves.* – In order to obtain the pressure-volume (or density)-energy relations for a medium from measurements of the kinematic parameters associated with mechanical shock-waves, it is desirable to make measurements in a one-dimensional, plane-flow system. As shown in Fig. 1 both flyer plate impact or the in-contact detonation of a high explosive will drive a plane-shock-wave into a sample. Much of the shock-wave data which have been obtained since 1955 have been obtained by either detonating explosives in contact with sample assemblies or using explosive detonation products to accelerate a metallic plate, or flyer plate (McQUEEN and MARSH [4]). The explosively driven plate, in turn, impacts the sample assembly. A clear advantage of the flyer plate impact method over the explosive in contact method is that the resulting pressure pulse delivered to the sample is flat-topped. The shock pulse induced by the flyer plate is steady for a time equal to the sum of the shock transit time through the sample plus the transit time of the wave, a rarefaction wave, which results from reflection of the shock in the flyer plate, at the rear surface of the flyer plate. In recent years, an increasing body of shock-wave data has been obtained below $\sim 50$ kbar, a regime where dynamic yielding effects in solids are important and interesting, by using a compressed

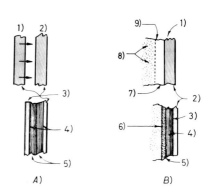

Fig. 1. – The generation of plane shock-waves: $A$) plate 1 impacts plate 2 and, as a result, shock fronts will propagate into both plates; 1), 2) plate; 3) initial-state $a$; 4) shock fronts; 5) shock-state $b$. $B$) Upon interaction of detonation wave at the explosive-plate 1 interface, forward shock propagates into plate 1 and the rearward shock propagates back into the detonation products: 1) plate; 2) initial-state $a$; 3) shock front in plate 1); 4) shock-state $b$; 5) explosive-plate 1 interface; 6) rearward traveling shock in detonation products; 7) undetonated explosive; 8) detonation products; 9) detonation front.

gas gun (THUNBORG *et al.* [9]). These devices accelerate a flyer plate bearing projectile to speeds of (1 to $\sim 1.5$) km/s. In 1966, JONES *et al.* [10] described the use of a two-stage light gas gun for accelerating flyer plate bearing projectiles to speeds of $\sim 7$ km/s. This latter technique, although requiring a large initial investment of equipment, is gaining rapid acceptance because of the inherently greater control which may be exercised over the shape and velocity of the flyer plates which are launched.

  **2'2.** *Calculating shock states.* – The thermodynamic states which are produced behind shock-waves are determined using the Rankine-Hugoniot conservation equations

$$\text{(1)} \qquad \varrho_b = \varrho_a (U - u_a)/(U - u_b) \, ,$$

$$\text{(2)} \qquad p_b = p_a + \varrho_a (U - u_a)(u_b - u_a)$$

$$\text{(3)} \qquad e_b = e_a + (p_a + p_b)/(1/\varrho_a - 1/\varrho_b)/2 \, ,$$

$$\text{(4)} \qquad h_b = h_a + (p_a - p_b)/(1/\varrho_a + 1/\varrho_b)/2 \, ,$$

which express conservation of mass (1), momentum (2), and internal energy (3), or enthalpy (4) across the pressure discontinuity or shock front. Here $\varrho$ is density, $U$ and $u$ are shock and particle velocity (with respect to the laboratory), $\varrho$ is shock pressure, or stress perpendicular to the plane of the shock, and $e$

and $h$ are specific internal energy and enthalpy. Subscripts $a$ and $b$ refer to parameters in front of, and behind, the shock front. It should be noted that the initial density of the unshocked material, $\varrho_a$, need not be the crystal density of the sample. By employing an artificially porous sample, with a $\varrho_a$ considerably lower than the crystal density, it follows from (3) that large increases in internal energy at a given compression may be obtained. Equations (1)-(4) simplify when the shock is incident upon material at ambient pressure and at rest since then, $u_a = p_a = 0$.

For strong shocks the mean, or hydrostatic, stress is assumed to be equal to the shock stress. At low-shock stresses yielding does not occur and the compression on a microscopic scale is uniaxial. In this case, for an isotropic solid the stress in the direction of shock propagation, $p_x$, is related to the stress transverse to the propagation direction, $p_y$, by

$$(5) \qquad\qquad p_x = (1 - \nu) p_y / \nu ,$$

where $\nu$ is the Poisson's ratio. The stress limit, $p_{HEL}$, for one-dimensional strain under shock loading is called the Hugoniot elastic limit ((HEL) Figure 2). The HEL is on the order of 0.1 kbar for alkali halides, 1 kbar for metals, and 50 kbar for silicates. Since states along the elastic Hugoniot, up to the Hugoniot elastic limit, are achieved via one-dimensional compression, the maximum shear stress, $\tau$, is given from elasticity theory by

$$(6) \qquad\qquad \tau = \tfrac{1}{2} (p_x - p_y) ,$$

or at the Hugoniot elastic limit $p_x = p_{HEL}$, it follows that

$$(7) \qquad\qquad \tau_{max} = [(1 - 2\nu)/(1 - \nu)][p_{HEL}/2] .$$

For shocks greater than the HEL, a range of pressures exists in which the final-shock pressure is achieved via two pressure discontinuities, or shock fronts; the first is called the elastic precursor. The second shock achieves the final state $(p_b, \varrho_b, u_b)$, which must be calculated using eqs. (1)-(4) with $p_a = p_{HEL}$ and $u_a = u_{HEL}$.

If the solid can retain some level of shear stress upon being shocked to states above the Hugoniot elastic limit, the resulting deformational Hugoniot curve will be offset above the « hydrostatic » curve by a stress $4\tau/3$ (Fig. 2). Here $\tau$ is the stress difference retainable upon dynamic yielding. For minerals such as quartz (WACKERLE [11]), periclase (AHRENS and LINDE [12]), and forsterite (AHRENS et al. [13]) comparison of the Hugoniot data upon being reduced to isothermal pressure-volume curves (see below), agree well with measured isotherms. On the other hand, Hugoniot data for other minerals such as calcite (AHRENS and GREGSON [14]) and sapphire (AHRENS *et al.* [15]) indicate that appreciable shear stresses are retained above the Hugoniot elastic limit.

Fig. 2. – The pressure-density plane showing the relation of the principal Hugoniot
to the adiabats and isotherms of the initial and high-pressure shock-induced phase.
Here $\varrho_{00}$, is the distended density of the
initial low-pressure phase, $\varrho_0$ the crystal
density, and $\varrho_0'$ the density at high temperature. $\varrho_0^h$ represents the zero-pressure
crystal-density of the shock-induced high-
pressure phase. — — — Principal adiabatic.
—·—·— 298 °K isotherm. — — — 0 °K isotherm. —·—·— High-temperature adiabatic.

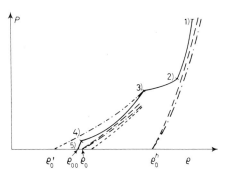

1), 2) Shock-induced phase; 2), 3) mixed-phase regime; 3), 4) initial phase;
4) Hugoniot elastic-limit; 4), 5) uniaxial strain.

2˙3. *Multiple shock fronts.* – As indicated in Fig. 2, a change in slope of the Hugoniot (see below) occurs at the HEL and at the onset and completion of shock-induced phase changes. In recent years, the search for, and detailed study of, the numerous shock-induced phase changes in elements and compounds has been on the forefront of research in the shock-waves in solids area. The discovery of the shock-induced $\alpha$ to $\varepsilon$ phase change in iron (BANCROFT *et al.* [16]), principally silicon to oxygen co-ordination changes in a large number of sili-cates and oxides (WACKERLE [11]; McQUEEN and MARSH [17]; AHRENS *et al.* [18]) have proved to be valuable in the interpretation of seismic data with regard to the study of the Earth's interior. As in the dynamic yielding (HEL) case, regions along the Hugoniot such as at the onset of phase change where $\partial^2 p/\partial v^2 < 0$ also give rise to multiple shock fronts.

The formation of multiple shock fronts for states just above a cusp or a region of anomalous curvature along the Hugoniot where $(\partial^2 p/\partial v^2)_{\mathrm{Hug}} < 0$, is illustrated in Fig. 3. Point $A$ may represent the onset of a shock-induced phase change from phase I) to phase II), or in the case of dynamic yielding the onset of the deformational regime where the local slope of the Hugoniot changes from

$$\left(\frac{\partial p}{\partial v}\right)_{\mathrm{Hug}} = -\,(K + 4\mu/3)/_{\mathrm{Hug}}V \quad \text{to} \quad \left(\frac{\partial p}{\partial v}\right)_{\mathrm{Hug}} = -\,K_{\mathrm{Hug}}/V.$$

Here $K$ and $\mu$ are the bulk and shear moduli of the solid as measured along the Hugoniot. For shocks driven into material, initially at state 0 ($p = 0$, $V = V_0$), achieving final states up to and including point $A$, a single shock whose velocity is given by

$$(8) \qquad U_1 = V_0 \left( \frac{p_1}{V_0 - V_1} \right)^{\frac{1}{2}}$$

is stable. Equation (8) follows from 1 and 2 using $1/\varrho = V$. We see that the propagation velocity of the shock is related to the slope of the line connecting point 0 to point $A$ in the pressure-volume plane. It follows from Fig. 3 that the velocity of a shock from a state 0 to state $C$, determined by the intersection of $0A$, with the upper portion of the Hugoniot curve, is given by

$$(9) \qquad U_3 = V_0 \left( \frac{p_3}{V_0 - V_3} \right)^{\frac{1}{2}} = U_1.$$

For a shock driven to some intermediate state between state $A$ and state $C$, such as state $D$, the shock velocity determined by the slope of a line (such as $AD$), will be less than that of a shock $0A$. Therefore, a faster shock, which has the velocity of $0A$, will form and compress the material from state 0 to state $A$; this will be followed by a second shock bringing the material from state $A$ to state $B$; this has a velocity

$$(10) \qquad U_2 = V_1 \left( \frac{p_2 - p_1}{V_1 - V_2} \right)^{\frac{1}{2}}.$$

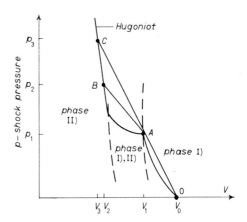

Fig. 3. – Shock pressure-volume Hugoniot delineating stability region for two shocks. Hugoniot curve $ABC$ intersects onset of phase transition, or Hugoniot elastic limit, at state $A$. Single shocks form upon achieving states 0 through $A$. States $A$ through $C$, such as state $B$, are achieved via two shocks. The first shock drives material from state 0 to state $A$, the second drives material from state $A$ to state $B$. For states above $C$ a single shock is stable.

The velocity of the first shock relative to the material behind it can be written using eqs. (1) and (2) as

$$(11) \qquad U_1' = V_1 \left( \frac{p_1}{V_0 - V_1} \right)^{\frac{1}{2}} = U_1 - u_1.$$

It is evident from eqs. (10) and (11) that the condition for the formation and stability of two shocks may be related to the slopes of the lines joining states $A$ and $B$, and $B$ and $C$, and is given by

$$(12) \qquad p_1/(V_0 - V_1) > (p_2 - p_1)/(V_1 - V_2).$$

Thus, $U_1 > U_2$, when eq. (12) is satisfied and for all shock states between $A$ and $C$, two shock fronts will form. States below $A$ and above $C$ are achieved via a single-shock front.

Some of the physical properties of solids when driven to high-pressure shock states which have been measured include electrical resistivity, dielectric constants, and absorption spectra. Most of the shock-wave measurements on geological materials have, to date, been directed toward determining various forms of the equations of state. By driving different amplitude shock-waves in a series of samples and measuring two kinematic variables, usually the shock velocity and flyer plate, or particle velocity, the thermodynamic state ($p_b$, $\varrho_b$, $e_b$) achieved behind the shock front is calculated from eqs. (1)-(4). A series of such thermodynamic states define the Hugoniot curve. If the initial material state is at ambient conditions, the principal Hugoniot curve is obtained. The principal adiabat is defined in an analogous way. The Hugoniot, being a locus of (shock) states, differs from other thermodynamic curves in that the thermodynamic path between the initial (state $a$) and shock (state $b$) states is not the Hugoniot curve itself but a straight (Rayleigh) line connecting states $a$ and $b$ (see for example, ZEL'DOVICH and RAIZER [19]).

## 3. – The complete equation of state.

In order to expand the knowledge of the complete equation of state, that is, to obtain the pressure-density-energy surface of a mineral, it is important to compare Hugoniot pressure-density curves with isothermal (constant, $T$) and adiabatic (constant entropy, $s$) curves. Isothermal and adiabatic curves are obtained from high-pressure X-ray and piston cylinder, and ultrasonic and Brillouin scattering experiments, respectively. Reduction of Hugoniot to adiabatic and isothermal conditions usually involves the Mie-Gruneisen equation

$$(13) \qquad\qquad p_h - p_k = \gamma \varrho (e_h - e_k) \,,$$

which relates the pressures and energies along the Hugoniot ($p_h$, $e_h$) with those along another thermodynamic curve at the same density, $\varrho$. The Gruneisen parameter, $\gamma$, is usually assumed to depend only on density and is given by

$$(14) \qquad\qquad \gamma = (\partial p / \partial E)_\varrho / \varrho = \alpha K_s / \varrho C_p \,,$$

where $\alpha$, $K_s$, and $C_p$ are

$$(15) \qquad\qquad -(\partial \varrho / \partial T)_p / \varrho \,, \qquad \varrho (\partial p / \partial \varrho)_s$$

and $(\partial e/\partial T)_p$. At a series of densities the pressure, $p_a$, along an adiabat centered at ambient pressure and a temperature $T_0$, are related to the pressure $p_h$ along the Hugoniot by

$$(16) \qquad p_h(1/\varrho_{00} - 1/\varrho)/2 = \int_0^{p_a} p' \, dp'/\varrho^2 + e_p + e_t + \int_{p_a}^{p_h} dp/\gamma\varrho \, ,$$

$$(17) \qquad e_p = \int_{T_{00}}^{T_0'} C_p \, dT \, ,$$

where in addition to the quantities defined in Fig. 2, $e_p$ and $e_t$ are the internal energy differences, if any, between the temperature and phase at which the Hugoniot is centered, $T_{00}$, and $e_a$, and that for which the adiabat is to be calculated. Integration paths for the first and last term are along the adiabat to be determined, and along the line, $\varrho = $ const. Supplementary shock-wave, thermodynamic data or solid-state theory must be used to supply $\gamma = \gamma(V)$ and $e_t$. The calculational method for obtaining isotherms and other forms of the equation of state are outlined in AHRENS et al. [18].

The application of shock data to initially porous materials for the purpose of determining the Gruneisen parameter at high pressure has been carried out by KORMER et al. [20] and McQUEEN et al. [21], for a series of metals. The method which has promise in determining the Gruneisen parameter of the high-pressure shock-induced phases (illustrated in Fig. 4), utilizes the Hugoniot curves of a series of initially (distended) samples of the subject material. Since at a given density in the high-pressure shock-state, the energy density induced by shock depends on the initial volume, $V_0 = 1/\varrho_0$, (see eq. (3)), and the initial volume can be extended to values considerably greater than the crystal volume, one can achieve different energy states at a constant density by driving shock-waves into materials with initially varying porosity. Given a set of such data, Gruneisen's ratios are then obtained by taking the finite difference

$$(18) \qquad \gamma \simeq V(\Delta p/\Delta e)_v \, .$$

Equation (18) assumes that the surface energy of the distended volume of the material, as well as any material strength which might prevent mechanical collapse of the pore structure, are negligible. If the differences in initial density arise because adjacent Hugoniot curves are centered on materials of different-initial phase, as for example, alpha-quartz and fused quartz, the difference in internal energy, $e_t$, must be taken into account. With this assumption the Gruneisen's ratio from eqs. (18) and (3) is

$$(19) \qquad \gamma = 2[\varrho(p_2/\varrho_{20} - p_1/\varrho_{10})/(p_2 - p_1) - 1 + \varrho e_t/(p_2 - p_1)]^{-1} \, .$$

As is shown schematically in Fig. 4, Hugoniot curves for materials which satisfy the Mie-Gruneisen formulation have an interesting property in that each curve, at sufficiently great compressions, will achieve a state at which

$$(20) \qquad (\partial p / \partial \varrho)_{\text{Hug}} \to \infty .$$

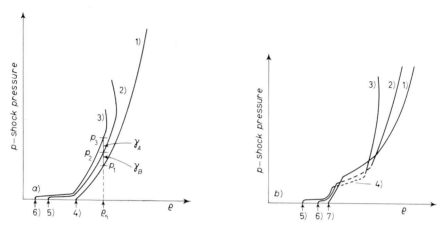

Fig. 4. – Hypothetical Hugoniot curves for initially porous materials *a*) Hugoniots for materials which remain in initial phase. Differences in pressure, $p_2-p_1$ and $p_3-p_1$, arise from differences in internal energies of materials shocked to same final density from various initial distentions: 1) single-crystal Hugoniot; 2), 3) porous-sample Hugoniot's; 4) single-crystal density; 5), 6) initially-porous samples; $\gamma_A \simeq (1/\varrho_h) \cdot$ $\cdot ((P_3-P_2)/(E_3-E_2))$, $\gamma_B \simeq (1/\varrho_h)((P_2-P_1)/(E_2-E_1))$. *b*) Hugoniot configurations for material with varying initial porosities, upon being shocked into a high-pressure phase regime: 1), 2), 3) high-pressure regime; 4) mixed-phase regime; 5), 6) low-pressure phase at various initial porosities; 7) single-crystal density.

At pressures higher than this state increasingly intense shock-waves will cause more thermal energy than lattice energy to be deposited in the material. Hence, at usually extreme pressures, the Hugoniot pressure becomes a double valued function of density. The density at which the condition (20) is satisfied may be obtained by substituting $e_b - e_a = \Delta e$ from eq. (3) into the Mie-Gruneisen equation to give

$$(21) \qquad p_h = p_k + \gamma \varrho (\Delta e - e_k) .$$

Here, $p_h$ and $p_k$ are the pressures along the Hugoniot, and along a reference curve which has internal energy $e_k$ at pressure $p_k$ and internal energy $e_a$, at $p_h = 0$. Upon differentiating (21) with respect to $\varrho$, it follows that

$$(22) \qquad \gamma|_{(\partial p_h / \partial \varrho)_{hu} = \infty} = 2/(\varrho/\varrho_0 - 1) .$$

Here $\varrho_0$ is the starting density of a material and, as already pointed out, this can, in general, be considerably lower than the single-crystal value. Thus, in addition to calculating a value of the Gruneisen parameter from the difference in pressure and energies along the Hugoniot curves, the Gruneisen parameter can also be calculated by using the overturning point of a single Hugoniot curve.

The values of Gruneisen's parameter obtained by the differencing method (eq. (18)), and from eq. (22) for stishovite are shown in Table I. As shown by the Table, there is a general tendency for the Gruneisen parameter to decrease

TABLE I. – *Gruneisen ratio for stishovite from porous-quartz and fused-quartz data.* (Modified from AHRENS et al. [22].)

| Sandstone [a], α quartz | | Fuzed quartz [b,c], α quartz | | Fused quartz [b,c], sandstone [a] | |
|---|---|---|---|---|---|
| $\varrho$, g/cm³ | $\gamma$ | $\varrho$, g/cm³ | $\gamma$ | $\varrho$, g/cm³ | $\gamma$ |
| 4.583 | 0.96 | 4.637 | 0.84 | 4.583 | 0.74 |
| 4.594 | 0.97 | 4.671 | 0.85 | 4.594 | 0.76 |
| 4.659 | 0.99 | 4.881 | 0.76 | 4.818 | 0.82 |
| 4.818 | 0.86 | 4.969 | 0.72 | 4.648 | 0.85 |
| 4.65 [d] | 1.49 [d] | 4.68 [e] | 0.88 [e] | 4.9 [d] | 1.63 [d] |
| 4.82 [d] | 1.40 [d] | | | 5.0 [d] | 1.57 [d] |

(a) Initial density, 1.98 g/cm³ (JONES et al. [25]).
(b) Initial density, 2.195 g/cm³ (SHIPMAN: private communication [58]).
(c) Data of SHIPMAN and WACKERLE [59] are combined.
(d) Limiting values from maximum density of Hugoniot curve.
(e) Average of 24 points calculated from data of McQUEEN [26].

with increasing density from the zero-pressure value of $1.6 \pm 0.4$ (AHRENS et al. [22]). Except for the $\gamma$'s determined from a differencing of the fused quartz and sandstone which presumably reflects the characteristics of a very much hotter, possibly molten, stishovite-type material, the values of $\gamma$ obtained from eq. (22) are on the order of a factor of 1.5 higher than those obtained from eq. (19). This probably reflects the fact that at sufficiently high energy densities the variation of $\gamma$ depends also on temperature.

The analysis for the equation of state of stishovite (density 4.28 g/cm³) which may be recovered for shock experiments on quartz (DECARLI and MILTON [23]) is significant because this mineral has a sixfold co-ordination of oxygen with silicon which is thought to be characteristic of silicates in the lower mantle of the earth. McQUEEN et al. [24] first recognized the high-pressure Hugoniot data of WACKERLE [11], obtained by shocking alpha-quartz and fused quartz, was representative of stishovite. Recently, thermal expansion and Hugoniot data in the stishovite regime for porous quartz (JONES et al. [25]) and new data for fuzed quartz (McQUEEN [26]; SHIPMAN and ISBELL, private communication) have combined to cal-culate the Gruneisen ratio at high pres-sure and the complete equation of state (AHRENS et al. [22]).

A reduction of the Hugoniot data for stishovite to a 25° isotherm as specified by the Birch-Murnaghan equa-tion, is shown in Fig. 5. This agrees closely with the high-pressure X-ray data to 200 kbar of LIU et al. [27]. A zero-pressure bulk modulus between 3.0 and 3.3 Mbar and $dK/dP$ of $\sim 7$ is ob-tained upon combining the shock- and high-pressure X-ray data.

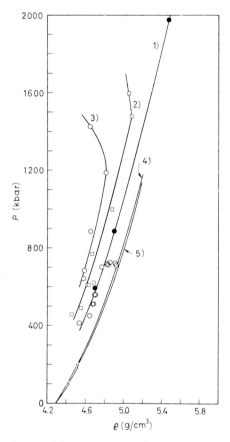

Fig. 5. – The pressure-density Hugoniot curves for solid and porous (sandstone) $\alpha$-quartz and fused quartz in the stishovite regime. The calculated adiabat for stishovite centered at standard conditions and the 25 °C isotherm for stishovite are calculated from the solid $\alpha$-quartz shock data. The isothermal X-ray data of LIU et al. and BASSETT and BARNETT for stishovite com-pares closely with the calculated 25 °C iso-therm. ○ WACKERLE [11]. ● AL'TSHULER et al. [60]. 1) Hugoniot, $\alpha$-quartz; 2) Hu-goniot fused-quartz; 3) Hugoniot coconino sandstone (JONES et al. [25]); 4) adiabatic stishovite; 5) 25° C isotherm stishovite; X-ray isotherms, stishovite: ▽ LIU et al. [27], △ BASSETT and BARNETT [61].

## 4. – Analytic forms for the equation of state.

Hugoniots, adiabats, and isotherms are usually specified by expressions containing two or more adjustable parameters. For raw Hugoniot data, a simple polynomial of the form

$$(23) \qquad U = c_0 + bu + cu^2 ,$$

is often used to describe the data. Introduction of eqs. (1) and (2) into (23) yields

$$(24) \qquad p_h = \frac{c_0^2 \varrho_0 y}{(1-by)^2} \left(1 + \frac{cp_h y}{2\varrho_0 c}\right)^2,$$

where $y = u/U$ and $c_0^2 = (\partial p_h/\partial \varrho)_s$, at $p_h = 0$.

Adiabats and isotherms are often expressed in the finite-strain Murnaghan and Birch-Murnaghan forms

$$(25) \qquad p = K_0(x^{3K'} - 1)/K'$$

and

$$(26) \qquad p = \tfrac{3}{2} K_0(x^7 - x^5)[1 - \xi(x^2 - 1)],$$

where $x = (\varrho/\varrho_0)^{\frac{1}{3}}$ and $K_0$ and $K'$ refer to the zero-pressure bulk modulus, $K' = (\partial K/\partial p)_s$ or $(\partial K/\partial p)_T$, at $p = 0$, and $\xi = 3(4 - K')/4$. The consistency of the Mie-Gruneisen assumption with various forms of the equation of state at high temperature is discussed by THOMSEN and ANDERSON [28].

The Birch-Murnaghan finite-strain equation is based on expanding the potential in terms of a strain to third order as measured from the final Eulerian configuration. In these proceedings THOMSEN has shown that by expanding the potential to fourth order in Lagrangian strain, in the case of a crystal with cubic symmetry, a complete equation of state based on the Mie-Gruneisen assumption which is thermodynamically self-consistent may be obtained (see also THOMSEN [29]). As is conventional, in formulating a complete equation of state, a static lattice potential at zero temperature is assumed. On the basis of the Lagrangian finite-strain theory, the lattice potential is of the form

$$(27) \qquad \varphi = \tfrac{9}{2} \tilde{K} \tilde{V} \eta^2 (1 - \Gamma\eta + \tfrac{3}{4}\Lambda\eta^2),$$

where the Lagrangian strain is

$$(28) \qquad \eta = \tfrac{1}{2} [(\tilde{V}/V)^{\frac{2}{3}} - 1].$$

Here $V$ is the specific volume of the lattice, and the quantities $\tilde{K}$ and $\tilde{V}$ are the hypothetical bulk modulus and specific volume at zero-temperature in the static lattice in the rest state, that is, properties of the solid that would exist if the material were lowered to 0 °K and the zero-point energy of the lattice removed. Here $\Gamma$ and $\Lambda$ are calculated from the ambient condition parameters (subscript, 0)

$$(29) \qquad \Gamma = \frac{K_0}{\tilde{K}}\left(\frac{\tilde{V}}{V_0}\right)\cdot\left(\frac{\partial K}{\partial p}\right)_{0T} + 3\Lambda\eta_0 ,$$

$$(30) \qquad \Lambda = \frac{K_0}{\tilde{K}}\left(\frac{\tilde{V}}{V_0}\right)^{\frac{5}{3}}\left\{K_0\left(\frac{\partial^2 K}{\partial P^2}\right)_{0T} + \left(\frac{\partial K}{\partial p}\right)_{0T}\left[\left(\frac{\partial K}{\partial p}\right)_{0T}+1\right]-\frac{1}{9}\right\}.$$

At any pressure and temperature, as well as along states along the Hugoniot, according to the thermal formulation of the Mie-Gruneisen equation of state, the pressures are given by

$$(31) \qquad p(V, T) = -\left.\frac{\partial\varphi_0}{\partial V}\right)_{T=0} + \frac{\gamma}{V}\, U_s(V, T) ,$$

where $U_s$ is the thermal energy which is given by

$$(32) \qquad U_s(V, T) = \tilde{U}_s(t)\cdot\left[1 - 3\tilde{\gamma}\eta\left(1 - T\,\frac{\tilde{C}_V}{\tilde{U}_s}\right)\right].$$

The terms $\tilde{U}_s(T)$ and $\tilde{C}_v$ refer to the internal energy and specific heat at constant volume in the rest state. The variation of thermal energy and the specific heat can be described by the Debye theory

$$(33) \qquad \tilde{U}_s(T) = n\tfrac{9}{8}\tilde{\theta}_D + \int_0^T D(\tilde{\theta}_D/T)\,\mathrm{d}T ,$$

where $\tilde{\theta}_D$, is the rest state Debye temperature, $D$ is the Debye function and $n$ is the number of atoms per molecule. Similarly,

$$(34) \qquad \tilde{C}_v(T) = D(\tilde{\theta}_D/T) .$$

The Gruneisen ratio in the rest state is determined from

$$(35) \qquad \tilde{\gamma} = \frac{\tilde{V}\alpha_0 k_0(V_0/\tilde{V})^{\frac{1}{3}}/\tilde{C}_{v0}}{\{1 + 3\eta_0[\lambda/\tilde{\gamma} + \tilde{\gamma}(\partial\ln\tilde{C}_V/\partial\ln T)_{T=T_0}]\}} ;$$

here $\alpha_0$ is the expansion coefficient at ambient conditions. The pressure at any point on the Hugoniot is then given by

$$(36) \qquad p_h(\eta, V_{00}) = \frac{[-\partial\varphi(\eta)/\partial p + (\gamma/V)(\tilde{V}_s - \varphi(\eta))]}{1 - \gamma(V_{00} - V)/2V} ;$$

here $V_{00}$ is the volume at zero-pressure at the foot of the Hugoniot, that is, the zero-pressure volume upon which that particular Hugoniot is centered. In this way we may represent the thermodynamic effect of initial porosity in the complete formulation of the Hugoniot. Also, $\gamma$ as a function of $V$ is obtained by expanding the thermal energy, $U_s(V, T)$ about the rest state with respect to $\eta$. This gives

$$(37) \qquad \gamma(V) = \left(\frac{V}{\tilde{V}}\right)^{\tfrac{4}{3}} (\tilde{\gamma} + 3\lambda\eta) .$$

The temperature, at any Hugoniot pressure, $P_h$, is readily calculated by noting that the shock pressure will agree with a particular isothermal pressure, for a temperature $T$, given by

$$(38) \qquad p_t = -3\tilde{K} \left(\frac{\tilde{V}}{V}\right)^{\tfrac{4}{3}} \left\{ \eta - \frac{3}{2}\eta^2 \Gamma + \frac{3}{2}\eta^3 \Lambda \right\} +$$

$$+ \frac{\tilde{U}_s}{\tilde{V}} \left(\frac{\tilde{V}}{V}\right)^{\tfrac{4}{3}} \left\{ \tilde{\gamma} + 3\eta \left[ \lambda - \tilde{\gamma}^2 \left( 1 - T\frac{\tilde{C}_V}{\tilde{U}_s} \right) \right] \right\} .$$

As an example of the use of these equations to the description of the Hugoniot of solids with different degrees of initial distension (porosity) the experimental Hugoniot data for MgO are shown in Fig. 5. The constants required in eq. (34) were obtained from fitting the ultrasonic data for single crystal MgO obtained to 1000 °K and 10 kbar by SPETZLER [30]. The constants for use in eq. (36) were specifically obtained by using Spetzler's data centered at 800 °K (summarized in Table II) because these data were thought to represent the material

TABLE II. – *Ultrasonic data and thermodynamic data* (SPETZLER [30]) *for single crystal* MgO.

| | |
|---|---|
| $T$ | $= 800$ °K |
| $V_{p=0}$ | $= 0.284\,503$ cm³/g |
| $C_p$ | $= 0.123\,51 \cdot 10^8$ erg/g °K |
| $\alpha$ | $= 0.431 \quad \cdot 10^{-4}$ (°K)$^{-1}$ |
| $K_T$ | $= 1.458\,8 \cdot 10^{12}$ dyne/cm² |
| $(dK_T/dP)_T =$ | $3.89$ |
| $(dK_T/dT)_P =$ | $-0.1999$ dyne/cm² °K |

at a temperature comparable to the Debye temperature. For a relatively incompressible material such as MgO, the quantity, $(d^2K/dp^2)_{0T}$, which is required in eq. (30) to calculate $\Lambda$, is not yet measurable with present ultrasonic techniques. In order to obtain a value for $(d^2K/dp^2)_{0T}$ a simple ionic central force potential of the form (Thomsen, to be published)

$$(39) \quad \varphi = \frac{Mq^2}{R} + \frac{1}{2} e(R - \tilde{R}) + \frac{1}{2!} f(R - \tilde{R})^2 + \frac{1}{3!} g(R - \tilde{R})^3 + \frac{1}{4!} h(R - \tilde{R})^4$$

is used to calculate the value of $(\mathrm{d}^2 K/\mathrm{d} p^2)_{0T} = 15.2$ (Mbar)$^{-1}$ for MgO at 800 °K. In the first, or Madelung term, $M$ is the Madelung constant, $q$, the electronic charge and $R$, the nearest neighbor distance. The next four terms in (39) represent the repulsive potential, the constants are evaluated in terms of the elastic constants and $\tilde{R}$ represents the nearest neighbor distance in the rest state. As can be seen from Fig. 6, the fourth-order anharmonic theory de-

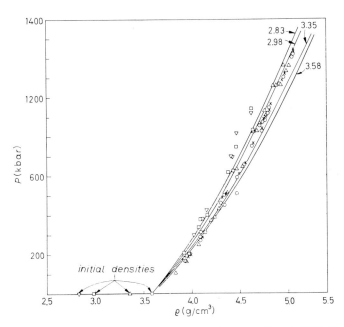

Fig. 6. – Hugoniot data for MgO which is shocked from various initial densities. At a given density, the less dense samples achieve, higher pressure, and hence, higher internal energies than the initial crystal density samples. Theoretical Hugoniots for various initial porosities based on Spetzler's [30] ultrasonic data on crystal MgO at 800 °K are indicated. MgO: $\varrho_0$ (g/cm³). Single crystal: 3.580, × McQueen and Marsh [17]; 3.585, ○ Carter et al. [62]. Porous: 3.350, △ Carter et al. [62]; 2.980, □ Carter et al. [62]; 2.830, ▽ Carter et al. [62].

scribes the thermal differences between the Hugoniots of the different initial porosity well. At a given density, the higher Hugoniot pressures for the more distended (less dense) initial samples reflect the higher thermal pressures. These results also demonstrate that there is significant divergence from the data at high compressions, $p_h > 400$ kbar, or, $V_0/V > \approx 1.2$. Here the fourth-order Lagrangian theory is inadequate and a higher order theory would be required to fit the Hugoniot curves.

## 5. – Summary of equation of state parameters for Earth materials.

Shock-wave data have been regularly collected in the Lawrence Radiation Laboratory Compendium (VAN THIEL et al. [31]). Parameters of the adiabatic equations of state and the temperature derivative of bulk modulus derived from shock data using the Birch-Murnaghan equation for several oxides are given in Table III. Adiabatic equations of state, also using the Birch-Murna-ghan equation for a series of shock-induced high-pressure phases of mixed oxides, silicates and rocks (in which the lattice co-ordination numbers are

TABLE III. – *Principal adiabats derived from shock data for minerals remaining in initial phase* (Birch-Murnaghan form, eq. (26)).

| Material | Bulk density (g/cm³) | STP density (g/cm³) | $K_s$ (Mbar) | $K_s'$ | $\gamma$ (a) | $(\partial K_s/\partial T)_p$ (kbar/°C) |
|---|---|---|---|---|---|---|
| MgO (single crystal) (b,f) | 3.584 | 3.584 | 1.60 | 4.16 | 1.6 $(\varrho_0/\varrho)^2$ | $-0.22$ |
| Al₂O₃ (ceramic) (c,f) | 3.829 | 3.987 | 2.50 | 4.11 | 1.4 $(\varrho_0/\varrho)^2$ | $-0.23$ |
| Al₂O₃ (crystal) (d,f) | 3.987 | 3.987 | 2.55 | 3.23 | 1.4 $(\varrho_0/\varrho)^2$ | $-0.17$ |
| MnO₂ (polycrystalline) | 4.31 | 5.29 | 3.40 | 1.67 | 0.67 (e) | — |
| Mg₂SiO₄ (polycrystalline) (g) | 3.10 | 3.223 | 1.29 | 6.0 | 1.17 | $-0.15$ |

(a) Dependence on density, assumed.
(b) Strength correction, $\sigma_h = 15(1 + 2.7\, p_h)$ kbar applied to raw data.
(c) Strength correction, $\sigma_h = 40(1 + 0.5\, p_h)$ kbar applied to raw data.
(d) Strength correction, $\sigma_h = 52(1 + 2.5\, p_h)$ kbar applied to raw data.
(e) Indicates Dugdale-McDonald formula used for volume dependence, from AHRENS et al. [18].
(f) DAVIES and AHRENS: unpublished.
(g) AHRENS et al. [13].

generally increased in the new, denser structure), are given in Table IV. In general, the density and structure of these new phases, which are believed to correspond to those stable at lower mantle pressures, are unknown (with the notable exception of stishovite, SiO₂). The zero-pressure density has been determined by constraining the density and seismic parameter, $\varphi_0 = K_s/\varrho_0$, to satisfy Anderson's [32, 33] seismic equation of state

$$(40) \qquad \varrho_0/\overline{M} = 0.0429\,\varphi_0^{\frac{1}{3}}.$$

TABLE IV. – *Principal adiabats for shock-induced high-pressure phases* [a] (Birch-Murnaghan form, eq. (26)).

| | Bulk density (g/cm³) | Mean atomic weight | Zero-pressure density (g/cm³) | $K_s$ (Mbar) | $K_s'$ |
|---|---|---|---|---|---|
| Twin sisters dunite | 3.32 | 20.9 | 3.94 | 2.15 | 3.45 |
| Hortonalite dunite | 3.79 | 25.1 | 4.59 | 2.37 | 3.14 |
| Forsterite, $Mg_2SiO_4$ | 3.05 | 20.1 | 4.18 | 3.19 | 2.83 |
| Fayalite, $Fe_2SiO_4$ | 4.25 | 29.1 | 5.03 | 2.17 | 2.81 |
| Hematite, $Fe_2O_3$ | 5.00 | 31.9 | 5.70 | 2.73 | 2.87 |
| Magnetite, $Fe_3O_4$ | 5.12 | 33.1 | 6.11 | 3.23 | 2.74 |
| Spinel, $MgAl_2O_4$ | 3.42 | 20.3 | 4.03 | 2.65 | 3.29 |
| Sillimanite, $Al_2SiO_5$ | 2.71 | 20.3 | 3.94 | 2.44 | 2.91 |
| Andalusite, $Al_2SiO_5$ | 3.08 | 20.3 | 3.84 | 2.19 | 3.19 |
| Bronzitite | 3.28 | 21.2 | 3.33 | 1.09 | 3.34 |
| Anorthosite | 2.73 | 21.0 | 3.57 | 1.48 | 3.11 |
| Oligoclase | 2.63 | 20.5 | 3.57 | 1.59 | 3.05 |
| Albitite | 2.61 | 20.4 | 3.69 | 1.84 | 3.05 |
| Microcline | 2.56 | 21.4 | 3.36 | 1.09 | 3.06 |
| Westerly granite | 2.63 | 20.6 | 3.90 | 2.22 | 3.00 |
| Olivinite | 3.31 | 21.2 | 4.28 | 2.96 | 3.35 |

(a) DAVIES and ANDERSON [37].

TABLE V. – *Shock-wave data relevant to the equation of state of the Earth's core* (Raw Hugoniots, eq. (24)).

| | $\varrho_0$ (initial density) (g/cm³) | $\gamma_0$ | $c_0$ (km/s) | $b$ | $c$ (s/km) |
|---|---|---|---|---|---|
| Iron | 7.85 | 1.69 | 3.574 | 1.920 | — 0.068 |
| Nickel | 8.874 | 1.93 | 4.602 | 1.437 | |
| Sulphur | 2.020 | | 3.223 | 0.959 | |
| Titanium | 4.528 | 1.09 | 4.877 | 1.049 | |
| Iron alloys   4%   Si | 7.646 | 2.12 | 4.072 | 1.563 | |
| 19.8% Si | 7.016 | 1.47 | 5.444 | 1.235 | |
| 10% Ni | 7.884 | — | 3.083 | 2.355 | — 0.1638 |
| 18% Ni | 7.96 | — | 3.717 | 1.904 | — 0.0731 |
| 26% Ni | 7.97 | — | 3.097 | 2.515 | — 0.2353 |
| Miscellaneous water | 0.998 | — | 1.647 | 1.921 | — 0.096 |

(a) Data collected from VAN THIEL [31] and McQUEEN et al. [21].

Several reductions of the Los Alamos high-pressure equation of state data on rocks and minerals have been published by ANDERSON and KANAMORI [34], AHRENS *et al.* [35], WANG [36], and more recently DAVIES and ANDERSON [37]. Shock-wave data for metals and alloys pertinent to the study of the Earth's core are summarized in Table V.

Two important results pertaining to the Earth's interior have been obtained, largely on the basis of shock wave equation of state data:

1) The Earth's mantle is composed primarily of ferromagnesium minerals such as $(Mg, Fe)_2SiO_4$ and $(Mg, Fe)SiO_3$ with a $Mg^{++}$ to $Fe^{++}$ ratio of $\sim 4$ (BIRCH [38]; ANDERSON [39]).

2) The seismic and shock-wave data for metals, when taken together indicate that the Earth's iron core is not primarily alloyed with Ni, as was previously inferred from meteorite compositions, but contains a substantial quantity, perhaps 20 %, of lighter elements such as Si, S, C, and Ti (McDONALD and KNOPOFF [40]; ANDERSON *et al.* [41]).

## 6. – Shock-induced phase changes in forsterite and enstatite. Some recent results.

Enstatite and forsterite represent the magnesium-end members of the orthopyroxene and olivine solid-solution (Fe-Mg) series, and are of special interest in the study of the Earth's mantle. The seismic velocity and density-depth profiles of the mantle, obtained from seismological studies, can be closely predicted using laboratory elastic-constant data by assuming that the rock present in the mantle is a suitable mixture of these two minerals with a gross Mg/Fe ratio of about 4 or 5 (GRAHAM [43]). Furthermore, the abrupt increases in both seismic velocity and density which are observed to begin at depths of about 400 and 650 km are consistent with a series of polymorphic-phase changes which are known to take place in rocks of olivine-pyroxene mineralogy (Fig. 7). At pressures below 150 kbar, or equivalently, to Earth depths of about 450 km, the stabilities of Mg-Fe pyroxene and olivines, as well as a series of structural analog compounds have been explored in some detail in the laboratory, see for example, RINGWOOD [44], AKIMOTO [45] and AKIMOTO and SYONO [46]. At higher pressures the nature of the phase transitions, that may take place in Mg-Fe silicates can be investigated using thermochemical calculations (as outlined below) and using data for analog compounds, *e.g.* germanates, and crystal-chemical systematics. At present, the only direct access to the density-pressure relations for Mg-Fe minerals at deep mantle pressures are the shock data themselves. Although useful in providing the density-pressure relations at high pressure, these data can usually only provide an upper limit to the phase transition pressures because of the often sluggish response of silicates to suddenly applied high pressures and the fact that relatively low temperatures are achieved for only microsecond time intervals upon shocking these relatively incompressible materials.

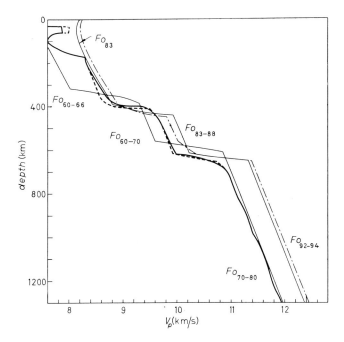

Fig. 7. – Two recent compressional velocity profiles for the upper mantle compared with velocities to be expected in olivine and peridotite mantles (modified from ANDERSON [39]). Olivine is assumed to undergo successive phase changes, first to the spinel structure and then to transform to a structure with the silicon six-fold co-ordination. Garnet is assumed to remain stable throughout the region shown. —————— olivine + pyroxene + garnet; — — — CIT 204M; ——— CIT 204.

In order to generate detailed equation of state data, delineating the phase-change region in forsterite and enstatite, a series of samples of these minerals were shocked to various pressures by impacting these with flyer plates as depicted in Fig. 1 A). The forsterite studied was stoichiometric, polycrystalline $Mg_2SiO_4$ cut from massive pieces which were fusion cast and had a bulk density of 3.1 g/cm³, as compared to crystal density of 3.22 g/cm³. Some pressed 2.6 g/cm³ samples were also studied. The enstatite studied (AHRENS and GAFFNEY [47]) was from Bamle, Norway, and is single crystal with a density of 3.29 g/cm³ and has a stoichiometry such that the Mg/Fe ratio is close to that of the Earth ... $Mg_{0.86}Fe_{0.14}SiO_3$. The flyer plates, contained in a series of plastic projectiles, were composed of a tungsten alloy (16.9 g/cm³, density). Impact

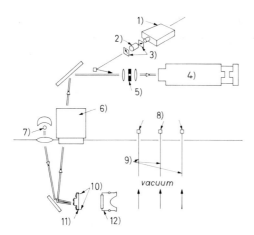

Fig. 8. – Diagrammatic sketch of an equation of state experiment after AHRENS *et al.* [13]. Projectile, which in our laboratory, is accelerated to ∼ 3 km/s by a propellent gun freely travels in vacuum. Upon cutting 3 laser beams and upon impact with pins mounted on the target, the velocity is determined. Successive interaction of the resulting shock in the sample with affixed mirrors, is recorded as a function of time with the streak camera; this determines the shock velocity in the sample. A Pockel's cell is used to intensity-modulate a pulsed argon laser; this provides 50 nanosecond timing marks for the streak camera which writes at speeds of 2 cm/μs on (10 × 20) cm film. 1) Pulsed argon laser; 2) Pockels cell; 3) polarizers; 4) image converter streak camera; 5) slit; 6) objective lens; 7) xenon flash tube; 8) photo diodes; 9) laser beams; 10) pins; 11) target; 12) projectile.

speeds in the range of 0.3 to 2.3 km/s were employed. The geometry used in the experiments and the resulting Hugoniot curves are depicted in Fig. 8-10.

Both enstatite and forsterite appear to undergo shock-induced phase changes. In the case of enstatite, work is still in progress. The results obtained to date indicate that this material displays a Hugoniot elastic limit of approximately 60 kbar and begins to undergo a phase change at about (130 ± 20) kbar. The

Fig. 9. – Hugoniot data for Bamle enstatite (after AHRENS and GAFFNEY [47]). Ultrasonic measurements of KUMAZAWA [48] and an assumed value of $K' = 5$ is used to construct theoretical principal isentrope for low-pressure (pyroxene) phase. Bamle enstatite (Mg 0.85, Fe 0.15) $SiO_3$: — — — Isentrope-chinoenstatite. 1) High-pressure regime; 2) mixed-phase regime; 3) low-pressure regime; 4) transition wave states 7.2 km/s; 5) Hel wave states 7.7 km/s.

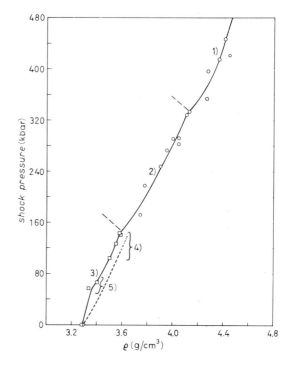

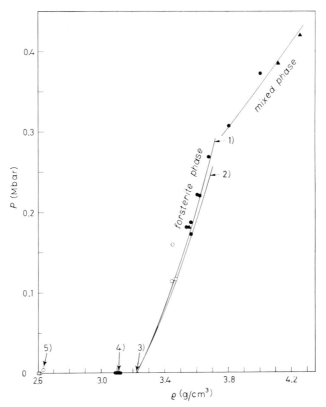

Fig. 10. – Hugoniot data for forsterite (after AHRENS *et al.* [13]). Birch-Murnaghan isentropes calculated from the Hugoniot data are in good agreement with those calculated from ultrasonic data. Initial density: o 2.6 g/cm³; • 3.1 g/cm³; ▲ 3.05 (McQUEEN and MARSH [17]). 1) Birch-Murnaghan (shock); 2) Birch-Murnaghan (ultrasonic); 3) crystal density; 4) initial-sample density; 5) Hugoniot-elastic limit.

elastic limit, and the somewhat higher pressure-induced phase-change causes a single shock to break up into multiple waves as discussed in previous Sections. The shock-wave speeds indicated, 7.7 and 7.2 km/s, represent the Hugoniot elastic limit, or precursor, shock and phase-transition shock, respectively. It appears (Fig. 9) that the Hugoniot of the low-pressure phase is offset above the insentrope inferred from the ultrasonic data of KUMAZAWA [48] indicating a strength effect of the type discussed earlier. Above 140 kbar the Hugoniot data indicate that states in a mixed phase or mixture of high-and low-pressure phase material are being achieved. Above 300 kbar five high-pressure Hugoniot states indicate that a material which has a zero-pressure density on the order of 3.7 g/cm³ is forming (*).

---

(*) Recent results are published in AHRENS and GOFFREY [63].

In the case of forsterite, states which are characteristic of forsterite, the low-pressure phase, are achieved at $(280 \pm 20)$ kbar. Above this pressure the Hugoniot also appears to enter a mixed-phase region. The high-pressure data obtained by McQUEEN and MARSH [17] and McQUEEN et al. [49], indicate that a high-pressure phase which has a density characteristic of mixed oxides ($\sim 4$ g/cm³) is forming. A series of possible high-density assemblages which may be forming under shock compression of olivine and pyroxene are listed in Table VI.

TABLE VI. – *Densities of high-pressure phases of magnesium-iron, olivine and pyroxene.*

| Structure | Density (g/cm³) | | |
| --- | --- | --- | --- |
| | $X = Fe^{++}$ | $X = Mg^{++}$ | $X = (Mg_{0.86}, Fe_{0.14})$ |
| $X_2SiO_4(\beta)$ | — | 3.47 [a] | |
| $X_2SiO_4(\gamma)$ | 4.85 [b] | 3.56 [a] | |
| $2\,XO + SiO_2$ (stishovite) | 5.36 | 3.85 | |
| $XSiO_3$ (ilmenite) | — | 3.80 [c] | 3.85 [d] |
| $\frac{1}{2}\,X_2SiO_4(\gamma) + \frac{1}{2}\,SiO_2$ (stishovite) | 4.68 | 3.75 | 3.89 [d] |
| $XSiO_3$ (perovskite) | | 3.95 [e]—4.25 [f] | |
| $XO + SiO_2$ (stishovite) | 5.04 | 3.97 | |
| Garnet structure | | | 3.67 [g] |

(a) RINGWOOD and MAJOR [51].
(b) MAO [53].
(c) RINGWOOD [44].
(d) Calculated from Bamle enstatite $Mg_{0.86}Fe_{0.14}SiO_3$:
(e) GAFFNEY and AHRENS [47].
(f) AHRENS et al. [18].
(g) Probable structure.

The nature of the high-pressure phases of dunites and bronzitites shocked by McQUEEN and his coworkers has already been discussed extensively by WANG [42], ANDERSON and KANAMORI [34], AHRENS et al. [18], and more recently by DAVIES and ANDERSON [37]. The significance of the present high-pressure phase transition data and their relevance to theoretical phase diagrams and the high-pressure phases which are present in the Earth are discussed in the next Section.

## 7. – Calculating phase diagrams from equation of state of thermochemical data.

In order to examine minimum pressures at a given temperature, either along the Hugoniot or in the Earth's mantle, at which a possible phase change is expected to take place, it is useful to construct theoretical phase diagrams

in $p$, $T$ and usually, binary chemical composition, space. Such theoretical con-
structions complement the high-pressure phase equilibria data which have
been obtained by RINGWOOD and AKIMOTO and their coworkers, and LINDSLEY
et al. [30] at comparable and lower pressures.

The phase diagram for $(Mg, Fe)_2SiO_4$ is known at temperatures from
$(600 \div 1000)$ °C from the detailed experimental investigations of AKIMOTO [45]
to below 100 kbar, and from the work of RINGWOOD and MAJOR [51] to over
200 kbar. Both groups have shown that for mole fractions of magnesium (Mg),
greater than 0.7, as are probably present in the mantle, transformation to a
$\beta$-$(Mg, Fe)_2SiO_4$ phase with a density which is $\sim 8\%$ greater than that of the
equivalent olivine composition, takes place below a depth of about 400 km.
At greater depths, further transformation to the $\gamma$-$(Mg, Fe)_2SiO_4$ phase, which
has the true spinel structure, takes place. This phase has a density which is
some 11% greater than the density of the equivalent olivine composition. The
similar reaction

(41)          $2(Mg, Fe)SiO_3 \rightarrow SiO_2(\text{stishovite}) + (Mg, Fe)_2SiO_4(\text{spinel})$

is predicted to take place below 400 km in the mantle for pyroxene compositions. At pressures of the order of 200 kbar (650 km deep in the mantle),
a further increase in density, as well as seismic velocity is observed (Fig. 7).
This increase corresponds to increases in density and velocity which can be
approximately predicted by assuming that a high-pressure phase is induced at
these pressures has properties similar to those of the mixed oxides MgO, FeO,
and $SiO_2$ (stishovite). It has been pointed out by RINGWOOD [44] that in the case
of olivine compositions this increase in density and velocity actually may
correspond to the formation of a $(Mg, Fe)_2SiO_4$ polymorph in the $Sr_2PbO_4$
structure. This is expected to have a similar density as the mixed oxides. The
pressures and temperatures required to induce the following reactions have
been examined for the case of Mg-Fe end members forming a series of solid
solutions: (In the case of disportionation, or, breakdown reactions, in order
to preserve the assumption of a two-phase system, the reaction products are
assumed to behave as a single phase, with thermochemical and equation of
state properties equal to the appropriate molar average.)

(42)      $X_2SiO_4$ (olivine) $\rightarrow 2XO$ (halite) $+ SiO_2$ (stishovite),

(43)      $X_2SiO_4$ (spinel) $\rightarrow 2XO$ (halite) $+ SiO_2$ (stishovite),

(44)      $2XSiO_3$ (clinopyroxene) $\rightarrow 2XO$ (halite) $+ _2SiO_2$(stishovite),

(45)      $2XSiO_3$ (clinopyroxene) $\rightarrow X_2SiO_4$ (spinel) $+ SiO_2$ (stishovite).

The ideality of solid solution assumption has been shown to be approximately valid in the case of the olivine-spinel transformations (AKIMOTO [45]) and by reaction (41), as demonstrated by the present calculational results.

In order to calculate the reaction pressures at various temperatures for the Mg-Fe end members of reactions listed above, it is convenient to employ standard enthalpies and entropies under standard temperature and pressure conditions for these compounds. For $Mg_2SiO_4$ and $Fe_2SiO_4$ in the olivine phase, $MgSiO_3$ (clinopyroxene), and for the oxides FeO, MgO and $SiO_2$ (stishovite) these data as well as the appropriate molar volumes are listed in Robie and Waldbaum's [52] recent compilation of thermochemical data. Thermochemical data for the spinels, $Fe_2SiO_4$ and $Mg_2SiO_4$, ($\gamma$-phase) have been calculated by MAO [53].

For $FeSiO_3$ (ferrosilite) the phase diagram for the reactions

(46)                $FeSiO_3$ (orthopyroxene) $\rightarrow FeSiO_3$ (clinopyroxene)

and the calculated phase line

(47)                $FeSiO_3$ (clinopyroxene) $\rightarrow Fe_2SiO_4$ (spinel) $+ SiO_2$ (coesite) ,

which have been studied by AKIMOTO et al. [54] were employed. The entropy and enthalpy of reaction for the clinoferrosilite to orthoclinosilite transition were used in calculating values for entropy and enthalpy of the clinoferrosilite. The resulting values were: 0.0217 kcal/mole °K and $-2.6315$ kcal/mole, respectively.

At a temperature $T$, the transformation pressure, $p_i^t$ for a given end member may be calculated from

(48)                $$\Delta G_i^{t,p} = \Delta H_i^t - T \Delta S_i^t = -\int_0^{p_i^t}(V_i^2 - V_i^1)\,dp \,.$$

Here $i$ indicates the end member (1 or 2) of a solid-solution series and $\Delta H_i^t$ and $\Delta S_i^t$ are the differences in enthalpy and entropy at zero-pressure and at temperature $T$. The integral in eq. (48) is along the isotherm at temperature $T$. Equation (48) must be solved for $p_i^t$ for each component, $i = 1, 2$, for each temperature. For a two-component, ideal mixing solid solution, system the mole fractions of component 1, in the low-pressure phase, $X_{11}$ in equilibrium with the high-pressure phase of composition $X_{21}$ are given by

(49)                $$X_{11} = \frac{\exp\left[\Delta G_1^{t,p}/nRT\right]\left(\exp\left[\Delta G_2^{t,p}/nRT\right]-1\right)}{\exp\left[\Delta G_2^{t,p}/nRT\right]-\exp\left[\Delta G_1^{t,p}/nRT\right]} \,,$$

(50)                $$X_{21} = \frac{\left(\exp\left[\Delta G_2^{t,p}/nRT\right]-1\right)}{\left(\exp\left[\Delta G_2^{t,p}/nRT\right]-\exp\left[\Delta G_1^{t,p}/nRT\right]\right)} \,,$$

where

(51)
$$\Delta G_i^{t,p} = -\int\limits_p^{p_i^t} (V_i^2 - V_i^1)\,dp\ .$$

Here $n$ is the number of atoms in each molecule which can enter the solid solution. The mole fraction of component two, in the low-pressure phase, $X_{12}$ is given by

(52)
$$X_{12} = 1 - X_{11}$$

and similarly the mole fraction of component two, in the high-pressure phase, is given by

(53)
$$X_{22} = 1 - X_{21}\ .$$

Equation (48) which determines the phase lines of the end members in $p, t$ space depend on knowledge of the enthalpies and entropies as functions of temperature and $p_i^t$ in eq. (48) and (51) depends on knowledge of the $P\text{-}V$ isotherms at various temperatures. A calculational method which has been used to determine these quantities is outlined below.

At each temperature it is assumed that the pressure-volume isotherms are given by the Birch-Murnaghan form of the isothermal equation of state (eq. (26)). It is also assumed that the thermodynamic quantity

(54)
$$\left(\frac{\partial p}{\partial e}\right)_v = \gamma/V$$

is constant for each material and has the value observed, or calculated, at standard temperature and pressure.

The temperature-dependence of $C_v$, the specific heat at constant volume, was assumed to be given by a Debye theory. The Debye temperature at each phase is estimated from its standard enthalpy, or, from specific-heat data, when these are available. To obtain an estimate of the variations in bulk modulus with temperature when this is not available experimentally, I have applied an empirical observation of ANDERSON [32]. In the case of oxides and silicates he observed that the variation of the bulk modulus with volume is similar when the volume change is produced by either compression or changes in temperature. Specifically, data for ten minerals listed by ANDERSON suggest the relation

(55)
$$\left(\frac{\partial \ln K_T}{\partial \ln \varrho}\right)_p \simeq \left(\frac{\partial \ln K_T}{\partial \ln \varrho}\right)_T + 0.85\ .$$

Taking eq. (55) as an equality it follows that

$$(56) \qquad \left(\frac{\partial K_T}{\partial T}\right)_p = -\alpha K_T \left[\left(\frac{\partial K_T}{\partial p}\right)_T + 0.85\right].$$

The properties of the various phases can be calculated numerically at a series of finite temperature increments of a function of temperature in the following order. The change in volume is calculated from the coefficient of expansion $\alpha$ The values of $\alpha$ are revised using the Gruneisen relation

$$(57) \qquad \alpha = \frac{\gamma}{V}\frac{C_p}{K_s}.$$

The change in the Debye temperature $\theta$ with changing volume is calculated from

$$(58) \qquad \theta = \theta_0 \exp\left[\gamma(V_0 - V)/V_0\right].$$

The latter equation may be derived from the fundamental definition of the Debye temperature and using the assumption of eq (54). A new value of the specific heat at constant pressure, $C_p$, is calculated from the thermodynamic identity

$$(59) \qquad C_p = C_v + \alpha^2 V T K_T$$

and the enthalpy and entropy change is calculated from

$$(60) \qquad \Delta H^t = \int_{T_0}^{T} C_p \, \mathrm{d}T,$$

$$(61) \qquad \Delta S^t = \int_{T_0}^{T} C_p \, \mathrm{d}T/T.$$

Using the above procedure, the theoretical phase diagrams calculated for reactions (42) to (45) are shown in Fig. 11-14. The direct transformation of olivine to post-spinel phase, represented by the mixed oxide calculation, takes place in the Earth via the $\beta$- and $\alpha$-(Mg, Fe)$_2$SiO$_4$ phases. In the case of shock experiments it is reasonable to assume a direct transition at pressures comparable to our results for pure forsterite. The onset of a shock-induced transition experimentally occurs at approximately $(280 \pm 20)$ kbar. A shock temperature of 460 °K was calculated at this point on the Hugoniot by AHRENS et al. [13]. If the high-pressure transition in Mg$_2$SiO$_4$, is indeed a phase with similar properties as those assumed for the oxide mixture, a transformation at

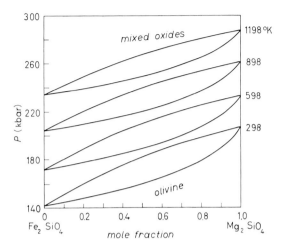

Fig. 11. – Theoretical-phase diagram for direct breakdown of (Mg-Fe) olivine to a polymorph which has the properties of the equivalent mixed-oxides (FeO, MgO and SiO$_2$ (stishovite)). A reaction to a high pressure phase of similar density is inferred to take place upon shock compression of olivine in the laboratory.

~ 210 kbar is obtained from the stability diagram of Fig. 11. The pressure for the transition for the Mg end member is somewhat lower than that observed in the shock case but considering the range of extrapolation of the thermo-

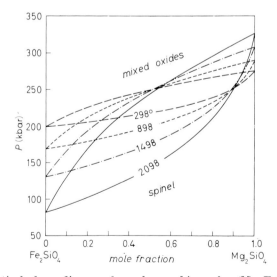

Fig. 12. – Theoretical-phase diagram for polymorphism of $\gamma$-(Mg, Fe)$_2$SiO$_4$ to a phase having similar properties as the mixed oxides. This hypothetical reaction, which has not been observed in either static or dynamic experiments, is believed to be related to the increase in density and seismic velocity in the mantle at depths of 650 to 700 km.

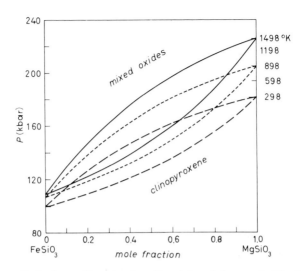

Fig. 13. – Theoretical-phase diagram for direct breakdown of (Mg, Fe) pyroxene to a phase having the properties of the mixed oxides. This reaction appears to take place at too high a pressure to account for the observed transition in the shock experiments on enstatite.

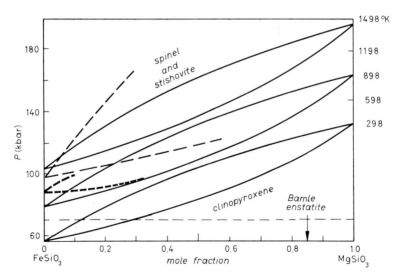

Fig. 14. – Theoretical-phase diagram for breakdown of (Mg, Fe) pyroxene to $\gamma$-(Mg, Fe)SiO$_4$+SiO$_2$ (stishovite). Comparison of theoretical diagram with experimental data of AKIMOTO and SYONO [46] indicate the approximate validity of ideal mixing assumption. This, or a related, reaction to the garnet structure may account for the phase change observed in the shock experiments on Bamle enstatite (Fig. 4). — — — Reconnaissance results $\sim$ 1300 °K, RINGWOOD and MAJOR [57]; – – – 1100 °K, AKIMOTO and SYONO [46];   —·—·— lower limit, stishovite stability (298 °K).

chemical data the agreement is remarkable; also, it is not impossible that in the dynamic situation, $\beta$- or $\gamma$-$Mg_2SiO_4$ forms first, and this in turn transforms to a high-pressure phase in the course of the shock experiment. Recently, BINNS [55] has reported the $\gamma(Mg, Fe)_2SiO_4$ phase, now called the mineral ringwoodite, in a highly-shocked zone in a meteorite.

The theoretical-phase diagram for the spinel to mixed oxides is shown in Fig. 12. A similar phase diagram was first calculated by ANDERSON [56]. This reaction which was thought to be similar to that reaction taking place at a depth $\sim 650$ km in the lower mantle in a material which has approximate stoichiometry of $(Mg_{0.6}, Fe_{0.4})_2SiO_4$ (ANDERSON [56]).

Interpretation of the enstatite results appears more straightforward. In this case the calculation indicates that it takes a greater shock pressure, at least 180 kbar at room temperature, than the 120 kbar observed to reduced the direct transformation of clinoestatites to mixed-oxide-type phase for a stoichiometry corresponding to Bamle enstatite. (The small density and enthalpy differences between the ortho- and clinopyroxenes are neglected.) In contrast to results for the transformation to the mixed-oxide-type phase, the calculated phase diagram for the reaction (41) (and (45)) gives a transition pressure of approximately 120 kbar. This lower limit is indicated in Fig. 14 and is close to the onset of the pressure stability range of stishovite. The agreement with Akimoto and Syono's [46] stability diagram near the $FeSiO_3$, stoichiometry is also indicated in Fig. 14. The agreement with their results as well as Ringwood and Major's [57] reconnaissance work, are remarkably good considering our present meager knowledge of the properties of ferrosilite. The relation of the transition pressure inferred from Fig. 14 to the results of Fig. 9 should, however, be considered as highly schematic, since it is quite possible that in the shock case the high-pressure phase that is forming is garnet structure which although according to Table VI has a zero-pressure density less than the mixture of $SiO_2$ (stishovite)+$(Mg, Fe)_2SiO_4$ (spinel) and probably differs in thermochemical properties.

\* \* \*

I am grateful for Leon THOMSEN's help in applying his finite strain formulation to calculating equations of state.

The experimental work which is summarized in this paper was carried out with the help of my colleagues, J. H. LOWER, S. GAFFNEY, and P. L. LAGUS. The support of the NSF for experimental work under GA 1650 and GA 21 396 and the theoretical studies under contract DA-49-146-XZ-089 and DASA(01-70-C-0021) is gratefully acknowledged.

# REFERENCES

[1] R. COURANT and K. O. FRIEDRICHS: *Supersonic Flow and Shock Waves* (New York, 1948).
[2] J. M. WALSH and R. H. CHRISTIAN: *Phys. Rev.*, **97**, 1544 (1955).
[3] J. M. WALSH, M. H. RICE, R. G. MCQUEEN and Y. L. YARGER: *Phys. Rev.*, **108**, 196 (1957).
[4] R. G. MCQUEEN and S. P. MARSH: *Journ. Appl. Phys.*, **31**, 1253 (1960).
[5] L. V. AL'TSHULER, K. K. KRUPNIKSOV, B. N. LEDENEV, V. I. ZHUCHIKHIN and M. I. BRAZHNIK: *Sov. Phys. JETP*, **34**, 606 (1958); L. V. AL'TSHULER, K. K. KRUKNIKOV and M. I. BRAZHNIK: *Sov. Phys. JETP*, **34**, 614 (1958).
[6] L. V. AL'TSHULER, A. A. BAKANOVA and R. F. TRUNIN: *Sov. Phys. JETP*, **15**, 65 (1962).
[7] M. VAN THIEL: *Compendium of shock wave data*, University of California, Radiation Laboratory Report, UCRL 50108 (1967).
[8] L. V. AL'TSHULER, B. N. MOISEEV, L. V. POPOV, G. U. SIMAKOV and R. F. TRUNIN: *Sov. Phys. JETP*, **27**, 420 (1968).
[9] S. THUNBORG jr., G. E. INGRAM and R. A. GRAHAM: *Rev. Sci. Inst.*, **35**, 11 (1964).
[10] A. H. JONES, W. M. ISABELL and C. J. MAIDEM: *Journ. Appl. Phys.*, **37**, 3493 (1966).
[11] J. WACKERLE: *Journ. Appl. Phys.*, **33**, 922 (1962).
[12] T. J. AHRENS and R. L. LINDE: *Response of brittle solids to shock compression, Behavior of Dense Media Under High-Dynamic Pressure* (New York, 1968), p. 325.
[13] T. J. AHRENS, J. L. LOWER and P. L. LAGUS: *Journ. Geophys. Res.*, **76**, 518 (1971).
[14] T. J. AHRENS and V. G. GREGSON jr.: *Journ. Geophys. Res.*, **71**, 4349 (1966).
[15] T. J. AHRENS, W. H. GUST and E. B. ROYCE: *Journ. Appl. Phys.*, **39**, 4610 (1968).
[16] D. BANCROFT, E. L. PETERSON and S. MINSHALL: *Journ. Appl. Phys.*, **27**, 291 (1956).
[17] R. G. MCQUEEN and H. M. MARSH: in *Handbook of Physical Constants*, edited by S. P. CLARK jr., Memoir 97, Geol. Soc. of America (1966).
[18] T. J. AHRENS, D. L. ANDERSON and A. E. RINGWOOD: *Rev. Geophys.*, **7**, 667 (1969).
[19] Y. B. ZEL'DOVICH and Y. P. RAIZER: *Physics of Shock-Wave and High-Temperature Hydrodynamic Phenomena*, Vol. **1** and **2** (English translation) (New York, 1967).
[20] S. B. KORMER, A. I. FUNTIKOV, V. D. URLIN and A. N. KOLESNIKOVA: *Sov. Phys. JETP*, **15**, 477 (1962).
[21] R. G. MCQUEEN, S. P. MARSH, J. W. TAYLOF, J. N. FRITZ and W. J. CARTER: *The equation of state of solids from shock-wave studies*, in *High-Velocity Impact Phenomena*, edited by R. KINSLOW (New York, 1970), p. 294.
[22] T. J. AHRENS and E. S. GAFFNEY: *Trans. Am. Geophys. Univ.*, **51**, 827 (1970); T. J. AHRENS, T. TAKAHASHI and G. DAVIES: *Journ. Geophys. Res.*, **75**, 310 (1970).
[23] P. S. DE CARLI and D. J. MILTON: *Science*, **147**, 144 (1965).
[24] R. G. MCQUEEN, J. N. FRITZ and S. P. MARSH: *Journ. Geophys. Res.*, **68**, 2319 (1963).
[25] A. A. JONES, W. M. ISABELL, F. H. SHIPMAN, R. D. PERKINS, S. J. GREEN and C. J. MAIDEN: *Material property measurements for selected materials*, in *General Motors Materials and Structures Laboratory Report*, NAS2-3427 (1968).
[26] R. G. MCQUEEN: *Shock-wave data and equation of state*, in *Seismic Coupling*, edited by G. SIMMONS (1968), p. 53.

[27] L. G. Liu, T. Takahashi and W. A. Bassett: *Trans. Amer. Geophys. Union*, **50**, 317 (1969).
[28] L. Thomsen and O. Anderson: *Journ. Geophys. Res.*, **74**, 981 (1969).
[29] L. Thomsen: *Journ. Phys. Chem. of Solids*, **31**, 2003 (1970); *The fourth order anharmonic theory, elasticity and stability*, to be published (1971).
[30] H. Spetzler: *Journ. Geophys. Res.*, **75**, 2073 (1970); L. Thomsen and O. L. Anderson: *Journ. Geophys. Res.*, **74**, 981 (1969).
[31] M. van Thiel: *Compendium of shock wave data*, University of California, Radiation Laboratory Report, UCRL 50108 (1967).
[32] D. L. Anderson: *Geophys. Journ.*, **13**, 9 (1967).
[33] D. L. Anderson: *Journ. Geophys. Res.*, **74**, 3857 (1969).
[34] D. L. Anderson and H. Kanamori: *Journ. Geophys. Res.*, **20**, 6477 (1968).
[35] T. J. Ahrens, D. L. Anderson and A. E. Ringwood: *Rev. Geophys.*, **7**, 667 (1969).
[36] C. Y. Wang: *Earth Planetary Sci. Lett.*, **3**, 107 (1967).
[37] G. Davies and D. L. Anderson: *Journ. Geophys. Res.*, **76**, 2617 (1971).
[38] F. Birch: *Journ. Geophys. Res.*, **69**, 4377 (1964).
[39] D. L. Anderson: *Mineral. Soc. Am.*, *Spec. Paper*, **3**, 85 (1970).
[40] G. J. F. MacDonald and L. Knopoff: *Geophys. Journ.*, **1**, 284 (1958).
[41] D. L. Anderson, C. Sammis and T. Jordan: *Composition of the mantle and the core*, in *Proceedings of the Francis Birch Symposium, April 1970* (in press).
[42] C. Wang: *Earth Planet. Sci. Lett.*, **3**, 107 (1967).
[43] E. K. Graham: *Geophys. Journ. Roy. Astr. Soc.*, **29**, 285 (1970).
[44] A. E. Ringwood: *Phys. Earth Planet. Interiors*, **3**, 109 (1970).
[45] S. Akimoto: *Phys. of Earth and Planet. Int.*, **3**, 189 (1970).
[46] S. Akimoto and Y. Syono: *Phys. Earth Planet. Int.*, **3**, 186 (1970).
[47] E. S. Gaffney and T. J. Ahrens: *Phys. Earth Planet. Interiors*, **3**, 205 (1970).
[48] M. Kumazawa: *Journ. Geophys. Res.*, **74**, 5973 (1969).
[49] R. G. McQueen, S. P. Marsh and J. N. Fritz: *Journ. Geophys. Res.*, **72**, 4999 (1967).
[50] D. H. Lindsley, B. T. C. Davies and I. D. MacGregor: *Science*, **144**, 73 (1964).
[51] A. E. Ringwood and A. Major: *Phys. Earth Planet. Interiors*, **3**, 89 (1970).
[52] R. A. Robie and D. R. Waldbaum: *Thermodynamic properties of minerals and related substances at 298.15 °K (25.0 °C) and one atmosphere (1.013 bars) pressure and at higher temperatures*, in *Geological Survey. Bullettin* (1968), p. 1259.
[53] H. Mao: *The pressure dependence of the lattice parameters and volume of ferromagnetism spinels, and its implications to the earth's mantle*, Ph. D. Thesis, University of Rochester (1967).
[54] S. Akimoto, T. Katsura, Y. Syono, H. Fujisawa and E. Komada: *Journ. Geophys. Res.*, **78**, 5269 (1965).
[55] R. A. Binns: *Phys. Earth Planet. Int.*, **3**, 156 (1970).
[56] D. L. Anderson: *Science*, **157**, 1165 (1967).
[57] A. E. Ringwood and A. Major: *Earth and Planet. Sci. Lett.*, **5**, 76 (1968).
[58] H. Shipman: private communication (1969).
[59] J. Wackerle: *Journ. Appl. Phys.*, **33**, 922, (1962).
[60] L. V. Alt'shuler, R. F. Trunin and G. V. Simakov: *Izv. Acad. Sci. USSR*, *Phys. Solid Earth*, **10**, 657 (1965).
[61] W. A. Bassett and J. D. Barnett: *Phys. Earth Planet. Int*, **3**, 54 (1970).
[62] W. J. Carter, S. P. Marsh, J. N. Fritz and R. G. McQueen: *The equation of state of selected materials for high-pressure references*, in *Accurate Characterization of the High-Pressure Environment*, edited by E. C. Hoyd (March, 1971), p. 177.
[63] T. J. Ahrens and E. S. Goffrey: *Journ. Geophys. Res.*, **76**, 5504 (1971).

# Shock Waves.

## Transport and Other Properties (*).

R. N. KEELER

*Department of Applied Science, University of California - Davis, Livermore, Cal.*
*Lawrence Radiation Laboratory, University of California - Livermore, Cal.*

Shock-wave techniques have been used for many years to provide data of geophysical interest on many minerals and those metals and alloys thought to make up the interior of the earth [1-3]. Certain inferences have been drawn from these studies. For example, from Hugoniot measurements by BALCHAN and COWAN [4] and limits established by BIRCH [5] from seismic data, it has been deduced that the earth's core is composed of an alloy of 80 % iron and 20 % silicon. In considering such arguments we should bear in mind that any candidate for a core material must satisfy not only equation-of-state and seismological data, but also criteria involving the earth's magnetic field, hydrodynamic flow within the core, and heat flow from the core to the mantle and thence to the earth's crust. Only recently have we had very high pressure measurements on properties of earth materials other than their equation of state. This is because of the nature of shock-wave experiments. Short times are always involved. The experiments are destructive in nature and, in general, it is difficult to make any but the simplest measurements.

Great attention has been given to possible origins of the earth's magnetic field [6]. The general consensus now is that this field is due to dynamo action in the liquid core, which apparently is precession-driven [7]. Solutions of the magnetohydrodynamic equations for dynamo action have proved quite difficult, and experimental verification of dynamo action in an experiment small enough for laboratory study seems nearly impossible, although some attempts have been made in this direction [8]. However, the vector equation

$$\frac{\partial \boldsymbol{B}}{\partial t} = \operatorname{curl}(\boldsymbol{u} \times \boldsymbol{B}) + (1/\mu\sigma)\nabla^2\boldsymbol{B}$$

(*) Work performed under the auspices of the U.S. Atomic Energy Commission.

can be nondimensionalized for study in terms of similarity considerations. When this is done, a nondimensional parameter is derived, the magnetic Reynolds number

$$R_m = 4\pi L V (\mu \sigma)$$

in which $L$ is a length scale associated with fluid motion in the core, $V$ is the time average core fluid velocity, $\mu$ is the magnetic susceptibility, $\sigma$ is the electrical conductivity, and $\mu\sigma$ is the inverse of the magnetic diffusivity.

It has been shown in shock-wave experiments that iron and iron-silicon alloys completely demagnetize at pressures above 0.5 Mbar [9]. Therefore, the magnetic susceptibility is unity, and ferromagnetism does not exist in either the solid or liquid part of the core (although some ferromagnetism may exist in the lower mantle) [10]. Observations of the drift of the earth's magnetic field and its symmetry indicate that convective motion in the core has a length scale of $10^3$ km and a drift velocity of about 10 km per year [11]. It is generally felt, for sustained dynamo action, that the magnetic Reynolds number must be above 100 [11]. This sets a lower limit on electrical conductivity at $3 \cdot 10^3$ mho/cm.

## 1. – Electrical conductivity.

The most extensive studies of this kind to date are high-pressure, high-temperature electrical conductivity measurements of iron, iron-nickel, and iron-silicon alloys. The techniques for making these measurements were discussed last summer in lectures [9] delivered at the International School of Physics « Enrico Fermi », entitled « The Physics of High Energy Density ». Some preliminary measurements were quoted. These measurements now are available in final form. Table I compares the conductivity of iron with that

TABLE I. – *Electrical conductivity of iron and iron alloys* (mho/cm).

| Material | Pressure | | |
| --- | --- | --- | --- |
| | 0.95 Mbar | 1.1 Mbar | 1.4 Mbar |
| Fe | $2.1 \cdot 10^4$ | $1.95 \cdot 10^4$ | $1.75 \cdot 10^4$ |
| Fe-20% Ni | | $1.07 \cdot 10^4$ | $1.06 \cdot 10^4$ |
| Fe-20% Si | $5.0 \cdot 10^3$ | $6.2 \cdot 10^3$ | $5.5 \cdot 10^3$ |

of iron-nickel and iron-silicon alloys at 1.4 Mbar, the pressure at the core mantle interface. It can be seen that the conductivity of the iron-silicon alloy is least able to sustain dynamo action.

It is instructive to consider some of the experimental limitations and difficulties encountered in shock-wave measurements of metallic conductivities. The metal sample, a thin rectangular strip, is sealed within a sandwich of insulating material. To reduce the effects of shock interaction, the dynamic impedance of the insulating material must be close to that of the sample and lead-ins. The insulator must remain an insulator at the high pressure and temperature attained behind the shock wave. No particular difficulties exist with the associated electronics; however, the circuits and oscilloscopes must be capable of resolving events taking place within times of 10 nanoseconds. The temperature of the shocked sample is calculated from the Grüneisen theory of solids in the Dugdale-McDonald approximation [12]. More sophisticated models [13] are available, based on perturbations around zero-pressure properties, but the application of these models is not warranted here in view of the extremes of pressure and temperatures involved.

We know from the seismology of the earth's core that the liquid portion is homogeneous. Because diffusive processes are quite slow, it follows that in shock-wave experiments we must start with a homogeneous material to get a homogeneous result. This has led to some difficulty, since many iron alloy compositions which would be desirable for experiments are in the metastable region at normal conditions. A small amout of inhomogeneity of sample, however, does not appear to affect conductivity results to any great extent.

Another question of concern is the state of the 80 % iron-20 % silicon alloy at the shock condition reached. We do not know whether it is liquid or solid. If the alloy is in the liquid phase, the 1.4-Mbar conditions very closely approach the state of the core at the core-mantle interface. However, if the results cited in Table I are for the solid phase, then the conductivity of liquid iron-silicon at 1.4 Mbar and the temperature required to liquefy it at that pressure will tend to make the true conductivity of the liquid phase even lower. The measured conductivity of 80 % iron-20 % silicon samples at 1.4 Mbar reported here may well prove to be an upper limit. If so, we have further evidence that the dynamo action of this proposed core material may be marginal.

I think it is clear that alloying materials other than silicon must be examined as core candidates. Perhaps sulfur or magnesium oxide [14] should be considered. In any case, the composition of the earth's core is still an open question, and the properties of the material finally selected must be consistent with seismic, thermal, and geomagnetic data.

## 2. – X-ray diffraction in shock-compressed solids.

One of the principal unsolved problems in the geophysics of the mantle is the nature of the material in the lower or the deep mantle. A number of phases have been proposed, most of them silicates with sixfold symmetry. Since the demarcation pressure between the upper and lower mantle is about 600 kbar, this region is not accessible to static experiments or static high-pressure crystallography. It would be of great interest to identify experimentally the high-pressure phases of the many minerals known to exist in the upper mantle and thus to gain a greater understanding of the properties of the lower mantle.

For the past two years JOHNSON and a group at the Lawrence Radiation Laboratory, Livermore, have been carrying out X-ray diffraction experiments on materials under shock compression. Some preliminary work was reported [14] at last summer's Enrico Fermi school, « The Physics of High Energy Density », describing some very low pressure (10-kbar) results obtained with a simple nondestructive shock-generating system. During the past year, X-ray diffraction has been observed in material under shock compression to 130 kbar [15]. The chosen sample was lithium fluoride. The 200 diffraction line was observed in the shocked state.

Several years ago, JOHNSON and others showed it possible to obtain a Debye-Scherrer pattern from a polycrystalline pressed-powder sample in less than 100 ns [16]. The X-rays were generated by large coaxial charged transmission lines capable of generating currents of more than 1000 A. The first set of shock-wave experiments was carried out at low pressure and was marginal for two reasons: nonhomogeneities of the resulting shock front, and difficulties encountered in timing the Blumlein flash X-ray generator [17]. These timing difficulties were overcome later by a modified X-ray trigger circuit which reduces jitter to less than $\pm 10$ ns.

Figure 1 shows the high-explosive shock-wave system most frequently used. A conventional plane-wave-generating high explosive provides the detonation wave for the plane-wave lens system. On striking the brass plate,

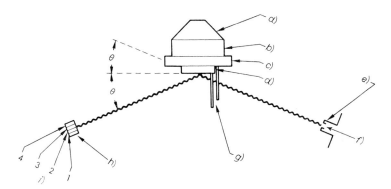

Fig. 1. – Sample geometry for X-ray diffraction study of LiF under shock compression. a) Plane-wave lens. b) TNT booster. c) Brass plate. d) LiF sample. e) Collimator. f) X-ray source. g) Shock pins. h) Scintillation detectors. i) Channel numbers.

the detonation wave becomes a shock wave and is driven through the lithium fluoride sample. Piezoelectric shock pins, positioned as shown, trigger the X-ray source and oscilloscopes. Another shock pin in the lithium fluoride sample provides interval time from which the shock velocity is calculated. A comparison of three experiments is shown in Table II. Experiments 1 and 3 were successful; experiment 2 was a failure, because of an error in timing the flash X-ray generator. The results from even the second experiment, however, verify the proper behaviour of X-ray intensities when the timing is premature.

TABLE II. – X-ray diffraction experimental data.

| Experiment | Detector | Intensity ratio | Timing |
|:---:|:---:|:---:|:---|
| 1 | 2<br>4 | 2.64<br>0.70 | Close to optimum: − 10 ns |
| 2 | 2<br>4 | 0.99<br>1.04 | Premature:        + 25 ns |
| 3 | 2<br>4 | 1.40<br>0 | Early:           + 10 ns |

From experiment 1, when the timing was very close to optimum, the 200 line of the shocked compressed state (detectors « 2 ») increased in intensity by a factor of 2.6, very close to the calculated value of 3.0. The 200 line of the unshocked state (detectors « 4 ») decreased to 0.7 of its predicted value. A moderate timing error was encountered in experiment 3, but when this is taken into account, the observed intensity agrees quite well with calculations.

These experiments clearly show the feasibility of carrying out X-ray diffraction experiments during high-explosive-driven shock compression of crystalline samples. Future plans include verification of the crystallographic phase changes observed in the alkali halides. Following this, experiments will be initiated on materials of geophysical interest.

REFERENCES

[1] L. V. ALT'SHULER: Sov. Phys. Uspekhi, 85, 9 (1965).
[2] R. G. McQUEEN and S. P. MARSH: Journ. Appl. Phys., 31, 1253 (1960).
[3] T. J. AHRENS and C. F. PETERSEN: Shock wave data and the study of the earth,

in *The Application of Modern Physics to the Earth and Planetary Interiors*, edited by S. K. RUNCORN, F.R.S. (New York, 1969), p. 449.

[4] A. S. BALACHAN and G. R. COWAN: *Journ. Geophys. Res.*, **71**, 3577 (1966).

[5] F. BIRCH: *Solids Under Pressure*, edited by W. PAUL and D. WARSCHAUER (New York, 1963), p. 137.

[6] For example, see J. G. TOUGH and R. D. GIBSON: *The Braginskii dynamo*, and following papers, in *The Application of Modern Physics to the Earth and Planetary Interiors*, edited by S. K. RUNCORN, F.R.S. (New York, 1969), p. 555.

[7] W. V. R. MALKUS: *Science*, **160**, 259 (1968).

[8] W. V. R. MALKUS: unpublished work.

[9] R. N. KEELER: *Proc. S.I.F.*, Course XLVIII (New York, 1971), p. 51.

[10] E. B. ROYCE: private communication.

[11] E. H. VESTINE *et al.*: *Journ. Geophys. Res.*, **72**, 4917 (1967).

[12] J. S. DUGDALE and D. MacDONALD: *Phys. Rev.*, **89**, 832 (1953).

[13] L. THOMSEN: *Journ. Phys. Chem. Solids*, **31**, 2003 (1970).

[14] B. J. ALDER: *Journ. Geophys. Res.*, **71**, 4973 (1966); see also D. L. ANDERSON, *et al.*: *Science*, **171**, 1103 (1971).

[15] Q. C. JOHNSON *et al.*: *Phys. Rev. Lett.*, **25**, 1099 (1970).

[16] Q. C. JOHNSON *et al.*: *Nature*, **5081**, 1114 (1967).

[17] Q. C. JOHNSON *et al.*: *Trans. Am. Crystallographic Ass.*, **5**, 133 (1969).

# Rheological Properties of the Deep Interior.

L. KNOPOFF

*University of California - Los Angeles, Cal.*

## 1. – Introduction.

Knowledge of the way in which mantle materials can deform under applied stress is essential to the construction of models of the mechanism for producing the motions observed in large scale plate-tectonics. The usual assumption in mathematical models is to take mantle materials as having a Newtonian viscosity. This is done for the obvious mathematical simplicity associated with the writing of the linear Navier-Stokes equation. What I propose to do in this lecture is to discuss the magnitudes of the physical properties of the mantle under long-term deformations as known at present, and to discuss what is known about the possible physical properties of materials at high pressures and temperatures, and hence to see if Newtonian viscosity is appropriate.

At present, there are significant differences of opinion whether or not it is indeed appropriate to describe the mantle of the Earth as having a Newtonian viscosity. We do know that this is completely inappropriate in the uppermost tens of kilometers of the Earth. In this region, earthquakes occur. Earthquakes are the often violent response of matter to external stress. At shallow depths, matter ruptures in brittle fracture along pre-existing, imperfectly healed fracture zones or faults. The great motions of the plates that cover the surface are probably rather uniform at the interiors near the regions remote from the places where earthquakes delineate the plate boundaries. We do not know this for a fact, but it seems plausible in view of the rather uniform rates of seafloor spreading given by the results from deep drilling at sea and from the results obtained from fission-track dating. In addition, earthquakes occur infrequently in plate interiors, implying small stress concentration gradients. But at the edges of the plates, where they are constrained by the friction against their neighbors, the motions are certainly irregular. Considerable pre- and post-shock creep may occur but it is doubtful, from laboratory experiments, that creep flow is Newtonian in character. The irregularity of the motions at the edges argues against a Newtonian process, for either the fracture or for the creep parts of the motions. Instead, both the creep and the fracture appear to be associated with the properties of materials at stresses close to their yield points; these stresses are not particularly low. In this case, the process must be nonlinear.

Fortunately, for my task, this school is devoted to the problems of the deeper mantle and core and hence I shall avoid the truly complex problems of the rheology of the outermost parts. Earthquakes extend to only a few tens of kilometers in depth, except at the places where the plates descend into the mantle and deep focus earthquakes are found. The near-surface regions are those of low pressure and temperature. Thus, our discussion centers on what happens to materials at high pressures and temperatures. I do not know of any materials that do not show a significant lowering of the fracture strength, of increased ductility and increased flow under laboratory conditions of high pressure and temperature. In the Earth, it is known that water, either in pore, interstitial or bound forms, acts as a significant agent in the determination of the strength of rocks. As water pressure is increased, strength decreases. We can say that probably the absence of earthquakes at depth, except for the regions near oceanic trenches, is due to the disappearance of fracture strength, due in turn to the increase of water pressure. In this state, the Earth must deform continuously by creep or plastic flow; this deformation is most likely nonlinear in character as is known from metallurgical experiments (OROWAN [1]). At these shallow depths, we are far from the melting point; however, it may be a dangerous extrapolation to assume the nonlinearity applies to the entire mantle.

Deep focus earthquakes appear to be a case of cold lithospheric matter being plunged into a hot environment. The consequences of the thermal disequilibrium are the earthquakes that are observed; the material moves in such a way to relieve these thermal stresses. The time constant for the process of reaching thermal equilibrium appears to be about 10 million years (ISACKS et al. [2]).

The mantle may well be close to its melting point over large parts of its interior, including at least the low-velocity channel under the oceans. It is, thus, useful to investigate the behavior of the long-period response to applied stress for materials close to their melting points. Before doing so, we review the measurements of « viscosities » for the Earth.

## 2. – Measurements of stress-strain rate properties of the Earth.

Most of the measurements of the long-period response of Earth materials to applied stress are interpreted in terms of a Newtonian viscosity. We write

$$\sigma = \eta \dot{\varepsilon} \ .$$

In general, the stress $\sigma$ and the strain rate $\dot{\varepsilon}$ are tensors of the second order and the viscosity $\eta$ is a tensor of the fourth order for isotropic Newtonian fluids. In this equation, the material is assumed to have zero strength and zero elasticity. However, we are mainly concerned with deformation in shear. Further, the earth is undoubtedly polycrystalline and the crystals are probably randomly oriented at sufficiently great depths in the Earth. Thus, we can consider $\eta$ as a scalar quantity, the shear viscosity. The kinematic viscosity $\nu$ is even more useful to discuss, especially since the dimensions of $\nu$ are those of any diffusion coefficient, cm²/s:

$$\nu = \eta/\varrho ,$$

with $\varrho$ the density.

Basalts have been tested in the laboratory at high temperatures for their values of $\nu$, but usually at temperatures above the solidus. These values of $\nu$ are, therefore, low compared to those for rocks below the solidus.

GRIGGS [3] measured $\nu$ for alabaster but at low temperatures and pressures. Equivalent values of $\nu \sim 10^{16}$ for this highly deformable rock were obtained under stresses of the order of hundreds of bars. A considerable number of measurements of $\dot{\varepsilon}(\sigma)$ relations, where $\dot{\varepsilon}$ is the strain rate and $\sigma$ is the applied stress, have been measured at $(p, T)$ conditions well below the melting point. Among the prominent names associated with this work are those of GRIGGS and BRACE and their colleagues. These results yield non-Newtonian $\dot{\varepsilon}(\sigma)$ responses, but this is to be expected.

In the field, the rebound of the surface of the Earth, after applied loads have been suddenly (on a geological scale of time) removed, has been interpreted to give Newtonian viscosities for the mantle. Two such measurements have been made, one for the removal of the Fennoscandian ice-cap and the other for the removal of the water load in Pleistocene Lake Bonneville. The crude result is the well-known number $\nu = 10^{21}$ cgs for the Fennoscandian case (HASKELL [4]) and the value $\nu = 10^{20}$ cgs for the case in western North America (CRITTENDEN [5, 6]). The contours of uplift for Fennoscandia have been interpreted by McCONNELL [7]. Of necessity, this inversion is highly nonunique, but it seems most likely that a low-viscosity layer is found under Fennoscandia between about 100 km to 400 km with $\nu \sim 10^{21}$ cgs and $\nu$ much larger both above and below the region. One can lower $\nu$ to about $3 \cdot 10^{20}$ cgs between 100 and 200 km with $\nu$ correspondingly higher between 200 and 400 km, if one wishes.

In the deeper mantle, estimates of the viscosity can be obtained from the earth's fossil bulge, assuming the stresses are supported by a viscoelastic structure (MUNK and MacDONALD [8]; McKENZIE [9]). The bulge is assumed to be an unrelaxed artifact of the time when the earth was rotating faster. For relaxation times of the order of 10 million years, $\nu$ for the lower mantle is about $10^{26}$ cgs.

McKENZIE [10] has sought to account for the bulge by assuming a poloidal convection presumably driven by the failure of the isotherms and the geoid to coincide, *i.e.*, the von Zeipel mechanism (EDDINGTON [11]). He finds this effect is too small by at least one order of magnitude to account for the observed bulge. Since the bulge is a term in harmonics of order 2, it cannot be generated by thermal convection driven from below. He concludes that a high viscosity in the lower mantle is required.

Recently, GOLDREICH and TOOMRE [12] have offered an alternate explanation for the bulge, stating that if the shape were expanded in harmonics about an earlier spin axis, the need for high viscosity in the lower mantle would disappear.

However, from arguments that state that the seismic transitions at about 400 and 600 km are phase transformations, one would expect a higher viscosity to be associated with a more compact crystal structure and hence a higher viscosity in the lower mantle compared with the upper. If these strong viscosity contrasts do indeed exist, the cause of drift of the upper part of the Earth must be found in the upper parts.

For the case of fluids heated from below, the horizontal wavelength of convection of the Rayleight-Bénard type is so much smaller than the distances continents have moved that one must search for sources in horizontal anomalies of the thermal state, *if* convection is confined to the upper mantle. Thus, the model that MALKUS has proposed is perhaps more to be favored than one of a layer driven by heating from below.

## 3. – Mechanism of creep at high temperatures.

It seems likely that the creep properties of solids are strongly temperature-dependent. Metals have been studied for some time because of their easy ductility but, until recently, the deformation of earthforming materials had not been studied. Experiments have now been performed on the properties of some few oxides at high temperatures and it is found that diffusion of lattice defects is important in determining the rheological properties.

Dislocations are relatively stationary at low temperatures but at somewhat higher temperatures, they become mobile and will diffuse to grain boundaries. The irregularity of grain boundaries, or rather their incompatibility with crystal structures, make them sources or sinks for these defects. This type of creep is a transient process; these dislocations are swept out of the system and are not present in steady-state or long-time constant processes (MOTT [13]).

Mobile interstitials and vacancies can also cause creep. Vacancies can be created at grain boundaries and moved to other grain boundaries under an applied external shear stress. In this case, creep exists in the steady state because the shear stress creates new vacancies continuously. The steady-state creep stress-strain rate relation is (HERRING [14]; NABARRO [15])

$$\dot{\varepsilon} = \frac{CDV_a}{kTa^2}\,\sigma\,,$$

where $D$ is the diffusion coefficient for vacancies, $C$ is a constant about 10 (MCKENZIE [16]), $V_a$ is the activation volume, $a$ is the grain radius and $kT$ is the thermal energy. From the temperature-dependence of the diffusion coefficient,

$$D = D_0 \exp\left[-(E + PV_a)/kT\right],$$

where $E$ is the activation energy. Thus, this type of creep yields an effective Newtonian solid with a viscosity

$$\eta = \frac{kTa^2}{CD_0V_a} \exp\left[(E + PV_a)/kT\right].$$

However, it is to be noted that this viscosity is extraordinarily tempera-ture-dependent as well as pressure-dependent. Considerations of the Navier-Stokes equations of motion in the presence of temperature and pressure gra-dients have not been made systematically as yet.

Grain boundary sliding is probably not important below about 10 to 15 km of depth in the Earth. Grain boundary sliding will only take on importance if one principal stress is negative and, at very small depths, the hydrostatic pressure soon exceeds the shear stress, making this process ineffective.

At low stresses but at relatively high temperatures, dislocations are rela-tively stable and act as sources and sinks of vacancies. But at higher stres-ses, the dislocations begin to move. At high temperatures, these dislocations can maintain their concentration by moving atoms through vacancy diffusion. WEERTMAN [17] has derived an expression for the creep rate in this case

$$\dot{\varepsilon} \sim (\sigma/\mu)^{4\frac{1}{2}}\,,$$

where $\mu$ is the shear modulus. It is of interest to note that, in the case of gla-ciers, which are surely solids close to their melting points and which exhibit considerable motion, NYE [18] and GLEN [19] have proposed the power law $\dot{\varepsilon} \sim \sigma^p$ with $p$ around 4.1 to explain the flow observed.

Thus, for the time being at least, the use of Newtonian viscosity to describe the rheological state of the interior is justified for those parts of the mantle under low shear stress, high confining pressure and close to the melting point. Diffusion creep is probably dominant at shear stresses less than about $10^{-5}\mu$ (McKenzie [16]).

## REFERENCES

[1] E. Orowan: *Phil. Trans. Roy. Soc.*, A **258**, 284 (1965).

[2] B. Isacks, J. Oliver and L. R. Sykes: *Journ. Geophys. Res.*, **73**, 5855 (1968).

[3] D. T. Griggs: *Bull. Geol. Soc. Amer.*, **51**, 1001 (1940).

[4] N. A. Haskell: *Amer. Journ. Sci.*, **33**, 22 (1937).

[5] M. D. Crittenden jr.: *Journ. Geophys. Res.*, **68**, 5517 (1963).

[6] M. D. Crittenden jr.: *Geophys. Journ. Roy. Astron. Soc.*, **14**, 261 (1967).

[7] R. K. McConnell jr.: *Journ. Geophys. Res.*, **70**, 5171 (1965).

[8] W. H. Munk and G. J. F. MacDonald: *The Rotation of the Earth* (Cambridge, 1960).

[9] D. P. McKenzie: *Journ. Geophys. Res.*, **71**, 3995 (1966).

[10] D. P. McKenzie: *Geophys. Journ. Roy. Astron. Soc.*, **15**, 457 (1968).

[11] A. S. Eddington: *The Internal Constitution of the Stars* (Cambridge, 1926).

[12] P. Goldreich and A. Toomre: *Journ. Geophys. Res.*, **74**, 2555 (1969).

[13] N. F. Mott: *Phil. Mag.*, **44**, 742 (1953).

[14] C. Herring: *Journ. Appl. Phys.*, **21**, 437 (1950).

[15] F. R. N. Nabarro: *Deformation of crystals by motion of single ions*, in *Conference on the Strength of Solids* (Bristol, 1948), p. 75.

[16] D. P. McKenzie: *The geophysical importance of high-temperature creep*, in *The History of the Earth's Crust*, edited by R. A. Phinney (Princeton, 1968), p. 28.

[17] J. Weertman: *Journ. Appl. Phys.*, **28**, 362 (1957).

[18] J. F. Nye: *Journ. of Glaciology*, **2**, 82 (1952).

[19] J. W. Glen: *Journ. of Glaciology*, **2**, 111 (1952).

# Pseudopotential Calculation of the Volume-Dependence of the Elastic Constants of Na at 0 °K.

C. M. BERTONI

*Istituto di Fisica dell'Università - Modena*

G. GIUNCHI

*Istituto di Geodesia dell'Università - Bologna*
*Istituto di Fisica dell'Università - Bologna*

## Introduction.

The pseudopotential theory has given a good description of the lattice dynamics of simple metals [1-2]. In this paper we want to use this theory to investigate the effect of a volume change, due to hydrostatic compression, on the frequencies of the normal modes of vibration in Na. In this way it is possible to obtain the volume-dependence of the elastic constants from which we may calculate the pressure-volume relation.

The comparison of our preliminary results with the experimental data at low temperature and high pressure shows the importance of considering the exchange and correlation correction in the dielectric description of the electron gas and the nonlocal features of the pseudopotential.

This theory of the equation of state at high pressure of the metals derives from first principles, but it is applicable only to the simple metals as alkali metals.

In the transition metals, the presence of the $d$-levels in the atoms makes the perturbation theory not applicable to the conduction electrons; at the moment however, it is possible to describe these metals only at high pressure with an approximate equation of state, like the Thomas-Fermi equation of state [3-4].

## 1. – Theory.

We describe metallic Na as a system of ions which, vibrating around the sites of a b.c.c. lattice, are embedded in an electron gas. In the « nearly free electron model » electrons are diffracted by a model pseudopotential due to the presence of the ions. We assume the energy-dependent model pseudopotential of Heine and Abarenkov [5-6]:

(1)
$$
\begin{cases}
v(r) = - \sum_l A_l(E) P_l = & r < R_M \\
\quad = - \dfrac{Ze^2}{r}, & r \geqslant R_M
\end{cases}
$$

where $P_l$ is the projection operator $|l\rangle\langle l|$ on the eigenstates of angular momentum $l$ and $E$ is the energy of the electronic state on which it operates.

Every $A_l$, that is the depth of the well of radius $R_M$ for each $l$, is fitted on spectroscopical data of the free ion.

$R_M$ is chosen to minimize the oscillations of the potential Fourier's transform $v(q)$ at large $q$.

In order to obtain dispersion relations of Na, we must find the eigenvalues of the dynamical matrix $D_{\alpha\beta}(q)$ through the equation [7]

(2)
$$
\sum_\beta \{ D_{\alpha\beta}(q) - \omega^2(q, \sigma) \delta_{\alpha\beta} \} e_\beta(q, \sigma) = 0 ,
$$

where $e(q, \sigma)$ is the eigenvector of the $q$ component of the ionic displacement with $\sigma$ polarisation.

The dynamical matrix is the sum of three contributions:

(3)
$$
D(q) = D^c(q) + D^e(q) + D^r(q) ,
$$

where $D^r(q)$ is the Born-Mayer repulsive term negligible in Na; $D^c(q)$ is due to the Coulomb interaction between ions [8] and $D^e(q)$ describes the indirect interaction between ions via the electron gas.

The pseudopotential (1) is screened by the « proton-electron dielectric function »:

(4)
$$
\varepsilon(q) = 1 + (1 - \mathscr{G}(q)) \frac{4\pi e^2}{q^2} \chi_0(q) ,
$$

$\chi_0(q)$ being the free-electron polarizability and $\mathscr{G}(q)$ the correction for exchange and correlation. When $\mathscr{G}(q)$ is omitted, one has the R.P.A. or Hartree approximation.

We have assumed for $\mathscr{G}(q)$ the interpolation formula given by SINGWI et al. [9]:

(5)
$$
\mathscr{G}(q) = A(r_s)[1 - \exp[-B(r_s) q^2/K_F^2]] ,
$$

where $r_s$ is the electron radius and $K_F$ Fermi's wavenumber.

The contribution to $D^e(\boldsymbol{q})$ of the second order in the pseudopotential is given by

$$(6) \quad D^e_{\alpha\beta}(\boldsymbol{q}) = -\Omega^2_p \left\{ \sum_{\boldsymbol{G}} \frac{(\boldsymbol{q}+\boldsymbol{G})_\alpha (\boldsymbol{q}+\boldsymbol{G})_\beta}{|\boldsymbol{q}+\boldsymbol{G}|^2} F_N(|\boldsymbol{q}+\boldsymbol{G}|) - \sum_{\boldsymbol{G}\neq 0} \frac{G_\alpha G_\beta}{|\boldsymbol{G}|^2} F_N(|\boldsymbol{G}|) \right\},$$

$$\Omega^2_p = \frac{4\pi e^2 Z^{*2}}{M\Omega_0},$$

where $\boldsymbol{G}$ are the reciprocal lattice vectors, $Z^*$ is the effective valence [10], $M$ the atomic mass and $\Omega_0$ the atomic volume.

$F_N(q)$ is the normalized « energy-wavenumber characteristic » [11] depending on the full nonlocal matrix element of the screened potential $\langle k+q|w|k\rangle$ given by ANIMALU [12].

The dispersion relation depends on the atomic volume not only through the factor $\Omega^2_p$, which appears also in the Coulombic part, but also from the screening and the potential.

Omitting nonlocality and energy-dependence, $F_N(q)$ is given by

$$(7) \qquad F_N(\boldsymbol{q}) = \frac{\Omega^2_0 q^4}{16\pi Z^2 e^2} |v^b(\boldsymbol{q})|^2 \frac{\varepsilon(\boldsymbol{q})-1}{\varepsilon(\boldsymbol{q})} \frac{1}{1-\mathscr{G}(\boldsymbol{q})},$$

where $v^b(q)$ is the local approximation of the bare potential.

From eq. (2) we can derive sound velocities

$$(8) \qquad u_\sigma(\boldsymbol{q}/q) = \frac{\mathrm{d}\omega(\boldsymbol{q},\sigma)}{\mathrm{d}q}\bigg|_{\varrho=0}.$$

A crystal with cubic symmetry has three independent elastic constants $c_{11}, c_{12}, c_{44}$ that, at 0 °K, in the framework of the harmonic approximation, are simply related to sound velocities [13]. We have determined them by velocities of sound waves travelling in the [110] direction through the relation:

$$(9) \qquad \begin{cases} c_{11} = \varrho(U^2_L - U^2_{T_2} + U^2_{T_1}), \\ c_{12} = \varrho(U^2_L - U^2_{T_2} - U^2_{T_1}), \\ c_{44} = \varrho U^2_{T_2}, \end{cases} \qquad \varrho = \frac{M}{\Omega_0}$$

the bulk modulus $B_0$ is also derived by

$$(10) \qquad B_0 = \frac{c_{11}+2c_{12}}{3}.$$

## 2. – Results.

In Table I are listed the values of the elastic constants and the bulk modulus which we have obtained in the local approximation of the potential and in the full nonlocal calculation, at the atomic volume of the uncompressed Na at 0 °K.

TABLE I. – *Elastic constants of uncompressed* Na (*units* $10^9 \, N \, m^{-2}$).

|  | Exp. (*) | Local potential | | Nonlocal potential | |
|---|---|---|---|---|---|
|  |  | ($^a$) | ($^b$) | ($^a$) | ($^b$) |
| $c'$ | 0.77 | 0.832 | 0.889 | 0.810 | 0.841 |
| $c_{11}$ | 8.9 | 17.71 | 11.61 | 17.24 | 9.77 |
| $c_{12}$ | 7.4 | 16.06 | 9.83 | 15.62 | 8.09 |
| $c_{44}$ | 6.6 | 6.67 | 6.79 | 6.95 | 7.05 |
| $B_0$ | 7.9 | 16.61 | 10.42 | 16.16 | 8.65 |

(*) Experimental data, athermal, from: R. H. MARTINSON: *Phys. Rev.*, **178**, 902 (1969).
($^a$) $\varepsilon(q)$ in the Hartree approximation.
($^b$) $\varepsilon(q)$ with exchange and correlation correction (5).

In both cases we present a comparison between the results derived in the Hartree approximation of $\varepsilon(q)$ and the results obtained including exchange and correlation correction (5) in the dielectric function. This describes correctly the electron gas compressibility [7] and so gives a better value of the deviation $(c_{12} - c_{44})$ from the Cauchy relation. Moreover, a correct calculation of the nonlocality and the energy-dependence of the potential improves these results and the shape of the dispersion relation, as it appears from Fig. 1, in which the obtained photon frequencies are compared with experimental data [14].

In Table II we give the values, which we have obtained with local calculations and the correction (5) to the dielectric function, of the elastic constants at different volumes of the elementary cell. To compare these results with experiments we have calculated a $P-V$ relation, at 0 °K, through the volume-dependence of the bulk modulus

(11)
$$P_0(V) = -\int_{V_0}^{V} \frac{B_0(V)}{V} \, dV .$$

These values are listed in Table III, in which are also reported, for comparison, the results obtained with Hartree's dielectric function.

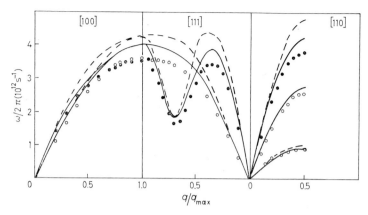

Fig. 1. – •, ○ Longitudinal and transversal data of 90 °K experiments of Woods *et al.* [10]. — — — Local potential approximation (with exchange and correlation term) ———— Nonlocal potential calculation (with exchange and correlation term). $q_{max}$ is the maximum value of the $q$-vector in the first Brillouin zone.

TABLE II. – *Volume-dependence of the elastic constants and bulk modulus of* Na, *calculated with local potential and exchange and correlation term* (5) (units $10^9 \, N \, m^{-2}$).

| $-\Delta V/V_0$ | 0 | 0.025 | 0.05 | 0.075 | 0.10 | 0.125 | 0.15 |
|---|---|---|---|---|---|---|---|
| $c'$ | 0.889 | 0.938 | 0.990 | 1.05 | 1.11 | 1.18 | 1.26 |
| $c_{11}$ | 11.61 | 12.49 | 13.45 | 14.49 | 15.63 | 16.87 | 18.23 |
| $c_{12}$ | 9.83 | 10.61 | 11.46 | 12.39 | 13.40 | 14.51 | 15.71 |
| $c_{44}$ | 6.79 | 7.07 | 7.36 | 7.66 | 7.97 | 8.29 | 8.62 |
| $B_0$ | 10.42 | 11.24 | 12.13 | 13.09 | 14.14 | 15.29 | 16.55 |

TABLE III. – *Pressure-volume relation of* Na.

| $-\Delta V/V_0$ | | 0.025 | 0.05 | 0.075 | 0.10 |
|---|---|---|---|---|---|
| Pressure ($10^4$ atm) | Exp. (*) (4.2 °K) | 0.19 | 0.41 | 0.65 | 0.93 |
| | Local pot. + Hart. app. | 0.43 | 0.88 | 1.37 | 1.88 |
| | Local pot. + ex. and corr. | 0.27 | 0.56 | 0.88 | 1.22 |

(*) Interpolated from data reported by: C. A. Swenson: *Phys. Rev.*, **99**, 423 (1955).

In conclusion we can say that the Hartree approximation of the screening gives bad results while an improvement is obtained by the use of Singwi's *et al.* dielectric function. The persistent deviation from the experiments is due to the local approximation.

Taking into account the non local pseudopotential, we have obtained a value of the bulk modulus of Na, at 0 °K and $P = 0$ volume, that is closer to the experiments (within 10%).

On the basis of this result we plan to extend the nonlocal calculation to higher pressures for Na and other simple metals.

The method we have presented points out that an accurate evaluation of the electron gas effects, and of the a model pseudopotential, gives, not only the correct dynamical properties of the metals, but also their volume-dependence.

* * *

We are grateful to Dr. V. BORTOLANI for having introduced us into this subject.

REFERENCES

[1] A. O. E. ANIMALU, F. BONSIGNORI and V. BORTOLANI: *Nuovo Cimento*, **44** B, 159 (1966).
[2] M. A. COULTHARD: *Journ. Phys. C: Solid St. Phys.*, **3**, 820 (1970).
[3] E. BOSCHI and M. CAPUTO: *Riv. Nuovo Cimento*, **1**, 441 (1969).
[4] E. BOSCHI and M. CAPUTO: *Lett. Nuovo Cimento*, **4**, 505 (1970).
[5] V. HEINE and I. ABARENKOV: *Phil. Mag.*, **9**, 451 (1964).
[6] A. O. E. ANIMALU and V. HEINE: *Phil. Mag.*, **12**, 1249 (1965).
[7] S. H. VOSKO, R. TAYLOR and H. KEECH: *Can. Journ. Phys.*, **43**, 1187 (1965)
[8] E. W. KELLERMANN: *Phil. Trans.*, A **238**, 513 (1956).
[9] K. S. SINGWI, A. SJÖLANDER, M. P. TOSI and R. H. LAND: *Phys. Rev. B*, **1**, 1044 (1970).
[10] R. W. SHOW and W. A. HARRISON: *Phys. Rev.*, **163**, 604 (1967).
[11] W. A. HARRISON: *Pseudopotentials in the Theory of Metals* (New York, 1966).
[12] A. O. E. ANIMALU: *Phil. Mag.*, **11**, 379 (1965).
[13] J. DE LAUNAY: *Solid State Phys.*, **2**, 219 (1956).
[14] A. D. B. WOODS, B. N. BROCKHOUSE, R. H. MARCH, A. T. STEWART and R. BOWERS: *Phys. Rev.*, **128**, 1112 (1962).

# Shell Structure on the Thomas-Fermi-Dirac Model for an Equation of State at Pressures of Geophysical Interest.

E. Boschi (*)

*Laboratoire des Hautes Pressions, C.N.R.S. - Paris*

M. Caputo

*Istituto di Fisica dell'Università - Bologna*
*Istituto di Geodesia dell'Università - Bologna*

It is generally believed [1] that the pressures in the Earth's core are too low for the applicability of simplified statistical treatments as the Thomas-Fermi-Dirac equation of state. Indeed TAKEUCHI and KANAMORI [2], interpreting the shock wave Hugoniot relations by means of a Mie-Grüneisen equation of state, have shown that the use of the Thomas-Fermi-Dirac model is not valid below a pressure of the order of 100 Mb. Nevertheless, an extensive use of the statistical theory of the atom, as worked out by THOMAS and FERMI and developed by DIRAC, WEIZSÄCKER and GOMBAS [3] has been made by a number of writers [4] to study the equation of state of the Earth's core and its chemical composition. This situation may be understood if we keep in mind that the Thomas-Fermi-Dirac model is the only physically acceptable approach in the high-pressure range. We are studying the possibility to use it in a more consistent way. The purpose of this intervention is to report on some preliminary results.

Our starting point is the obvious observation that the model, in its original formulation, cannot give account of the existence of discrete energy levels for bound electrons or, in other words, is not sensitive to the effects of shell structure. It is also obvious that these effects are the principal cause of the inapplicability of the Thomas-Fermi-Dirac model in the pressure range of geophysical interest.

---

(*) On leave of absence from Istituto di Fisica, Istituto di Geodesia, Università degli Studi, Bologna.

Until today, these effects have been considered only from a qualitative point of view and the influence of them on the equation of state has not been de-determined. So the first step is to find a way to introduce the shell structure in the theory. We believe that it is possible to solve consistently this difficulty by means of an appropriate modification of the normal ionization of the Thomas-Fermi-Dirac electronic charge cloud. As is well-known, in the Thomas-Fermi-Dirac theory, the number of the possible states per volume unit of the electrons, whose momentum is between $p$ and $p + dp$, is $8\pi p^2 dp/h^3$. Their density is given by the Fermi-Dirac statistics:

$$(1) \qquad d\varrho = \frac{8\pi p^2 \, dp}{h^3 \exp\left[(E_{\text{kin}} - e\Phi)/kT + \mu\right] + 1},$$

where $E_{\text{kin}}$ is the kinetic energy, $\Phi$ the Thomas-Fermi potential, $k$ the Boltzmann constant, $T$ the temperature and $\mu$ the chemical potential. If we consider an atom of atomic number $Z$, the integral of (1) over the entire phase-space atomic volume must satisfy the condition

$$(2) \qquad \frac{32\pi^2}{h^3} \int_0^R \int_0^\infty \frac{r^2 \, dr \, p^2 \, dp}{\exp\left[(E_{\text{kin}} - e\Phi)/kT + \mu\right] + 1} = Z.$$

Furthermore, Poisson's equation gives another relation between the electron density and the potential.

Now we assume that only a part of the electrons in the atom can satisfy the Thomas-Fermi-Dirac requirements. The assumption means that there is, for a given value of the pressure and of the temperature, some « critical radius » $R_c$ which divides the atom volume into two parts: an outer part containing electrons which can be actually considered as a charge cloud, and an inner part containing electrons whose behaviour needs a quantum-mechanical description. So, for given external conditions, we have $B$ « bound » electrons and $F$ « free » electrons, where $B$ and $F$ are linked by the obvious relation: $F + B = Z$. In a more precise approach we introduce a cut-off in the momentum phase space of the atom, that means change the lower limit of momentum integration, 0, which appears in eq. (2) to some value $p_c$. It is also necessary to find an appropriate potential to substitute $\Phi$ in (2) and to use it in the Schrödinger equation for the calculation of the number of bound electrons. This is the second step.

From the precedent discussion it seems to us natural to keep in mind the ionization phenomena and to represent the passage from « bound » electrons to « free » electrons as some kind of ionization induced by pressure. This makes possible the task of finding an appropriate potential function. To perform the calculations, we can follow two ways; the former consists in linking the two parameters of the problem, $R_c$ (or $p_c$) and $B/Z$ (or $F/Z$) to two measurable quantities as, for instance, the compressibility and its derivative; the latter consists in an iterative process starting from a trial value of the electron density. For the moment we have performed only the latter for an atom of atomic number 26, the most interesting in the Earth's core.

So we have found a great discrepancy with the Thomas-Fermi-Dirac pure model in the pressure range $(0 \div 100)$ Mb and, practically, the same asymptotic behaviour, confirming the observation of Takeuchi and Kanamori, and our recent work [5] in the range $(0.1 \div 5)$ Mb. Another interesting result is that above 0.5 Mb we have a large increase of density over a very small variation of pressure. This could become clear by a better understanding of the pressure ionization mechanism here introduced. Here we only want to outline that the Thomas-Fermi-Dirac is still the best approach to the Earth's core equation of state.

In our previous paper, ref. [5], the model of the atom under pressure was obtained (in the pressure range of geophysical interest), at 0 °K, starting from the observation that the pressure given by the Thomas-Fermi model can be considered as the incompressibility, and developed on the basis of the kinetic theory of gases.

The corrections to the Thomas-Fermi equation of state are in the same direction in both cases, in the present approach and in that of ref. [5].

## REFERENCES

[1] See, for example: J. A. JACOBS: *The physical properties of the earth's mantle and core*, this volume, p. 65.

[2] H. TAKEUCHI and H. KANAMORI: *Journ. Geophys. Res.*, **71**, 3985 (1966).

[3] See, for example: R. P. FEYNMAN, N. METROPOLIS and E. TELLER: *Phys. Rev.*, **75**, 1561 (1949); J. J. GILVERRY: *Equations of state at high pressure from the Thomas-Fermi atom model*, in *The Application of Modern Physics to the Earth and Planetary Interiors*, edited by S. K. RUNCORN (London, 1969); E. BOSCHI: *Equazioni di stato della materia per alti valori della densità e della temperatura*, Thesis, Scuola di Perfezionamento in Fisica, Bologna (1970).

[4] H. JENSEN: *Zeits. Phys.*, **111**, 373 (1938); W. M. ELSASSER: *Science*, **113**, 105 (1951); F. BIRCH: *Journ. Geophys. Res.*, **57**, 227 (1952); L. KNOPOFF and G. J. F. MACDONALD: *Geophys. Journ.*, **3**, 68 (1960); E. BOSCHI and M. CAPUTO: *Rivista del Nuovo Cimento*, **1**, 441 (1969).

[5] E. BOSCHI and M. CAPUTO: *Lett. Nuovo Cimento*, **11**, 505 (1970); *Annali di Geofisica*, **23**, 159 (1970).

# INVERSE PROBLEMS

## An Introduction to Earth Structure and Seismotectonics.

F. PRESS

*Earth and Planetary Science Department, M.I.T. - Cambridge, Mass.*

## 1. – Introduction.

It is impossible to summarize adequately the major portion of solid-earth geophysics in three lectures. This will be a feeble attempt to at least introduce the subject by explaining the motivations, describing the techniques and giving some of the results. Later lectures in the course will give the more advanced and specialized material necessary for full appreciation and comprehension.

The main goal of geophysics is to infer the properties of the Earth's interior in terms of composition, physical state, dynamic processes which occur today and which in the past have been responsible for the evolution of the Earth to its present form.

The principal methods for attacking this problem may be summarized as follows:

1) *Astronomical observations*:

   *a*) orbital-mechanical data to obtain mass $M$ and moment of inertia $I$.

   *b*) astrophysical and geochemical observations which provide the relative abundance of elements in stars, meteorites, planetary atmospheres.

2) *Geophysical experiments*. Conducted at or near the surface of the Earth to obtain the variation of elastic, electrical thermal and magnetic properties with depth.

   *a*) *Seismological observations*. Provide information on an important manifestation of dynamic processes—earthquakes; seismic waves from earthquakes and explosions reveal elastic and density layering within the interior.

   *b*) *Magnetic observations*. The Earth's field, its secular variation and its reversals provide clues to the physical state of the core. The history of the field as evidenced through fossil magnetism and magnetic imprinting of ancient rocks is a major element in understanding the mobility of the Earth's lithosphere.

    *c) Gravitational observations.* Anomalies in the gravity field are related to lateral density variations which are interpreted as variations in the thickness of subsurface layers and also as indications of the Earth's ability to support stress differences—*i.e.* the strength of the crust and mantle.

    *d) Thermal observations.* The temperature gradient near the surface together with the heat flow are boundary conditions for deducing the temperature distribution in the earth and its thermal evolution. Also needed is information on the distribution of radiogenic elements, melting points, thermal constants.

    *e) Geological observations.* Provide boundary conditions in terms of initial composition, physical properties and surface manifestation of dynamic properties such as volcanism, growth and movement of continents and xenoliths from the upper mantle.

    *f) Deep drilling.* A new tool for sea floor exploration which has verified the hypothesis of sea floor spreading and continental drift and provided the precise chronology (or rates) for the mobility of the outer layers of the Earth.

    *g) Laboratory experiments.* In which physical properties of crustal rocks, candidate rocks for the mantle, meteorites and moon rocks are determined under pressure and temperature conditions of the interior. More specifically, elastic velocity, density, thermal constants, strength, electrical properties and phase changes can be determined to depth equivalents of several hundred kilometres on Earth and the center of the moon. Using shock waves some of these properties can be inferred for pressures and temperatures equivalent to the center of the Earth.

In what follows I will draw primarily on 2 *a*), *b*), *e*), *f*), *g*), to provide preliminary notions of Earth structure and processes.

Specialized lectures on these and other topics are scheduled later in the course.

## 2. – Seismic waves (Ref. [26, 27]).

It can be shown that the equations which govern elastic wave propagation in a solid are

$$(1) \qquad \frac{\partial^2 \theta}{\partial t^2} = \alpha^2 \nabla^2 \theta \,,$$

$$(2) \qquad \frac{\partial^2 \Omega}{\partial t^2} = \beta^2 \nabla^2 \Omega \,,$$

where $\theta$ is a dilatation and $\Omega$ is a rotation. The two coefficients $\alpha$ and $\beta$,

$$(3) \qquad \alpha = \sqrt{\frac{k + \frac{4}{3}\mu}{\varrho}} \, ,$$

$$(4) \qquad \beta = \sqrt{\frac{\mu}{\varrho}} \, ,$$

give the velocity of compressional and shear waves. These are the $P$ and $S$ waves of seismology.

*Body waves* are $P$ and $S$ waves propagating in the Earth's interior. They are reflected and refracted where velocity and/or density changes occur. Body waves are excited by explosions or earthquakes. The amplitude, polarity and spectrum of body waves provide information on the mechanics of earthquakes and explosions, and the precise travel times of these waves reveal elastic layering and the variation of elastic constants with depth in the Earth.

Some examples of the use of body waves:

*Refraction experiments.* Figure 1 shows the principle and Figure 2 shows an actual ocean refraction profile. This method is used in oil-exploration and also to map the thickness of the Earth's crust. A major result of refraction shooting is the demonstration that the crust is thin under oceans (5 km) and thick under continents (35 km) verifying the isostatic concept of « floating continents » (Figure 3).

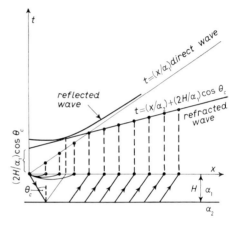

Fig. 1. – Schematic representation of reflected and refracted compressional waves.

*Body wave travel times.* Figure 4 shows theoretical trajectories for $P$ waves in the interior according to two different models of $\alpha$ *vs.* depth. The lower model is closer to reality. It shows the effects of a crust, a low-velocity zone and 2 transitions to higher velocities. These conclusions are reached from an analysis of travel times and spatial derivatives of travel times. A famous equation in seismology is the Herglotz-Wiechert-Bateman equation to invert body-wave travel times to obtain $\alpha(r)$, $r$ is distance from Earth's center:

$$(5) \qquad \int_0^{\Delta_1} \cosh^{-1}\left(\frac{p}{p_1}\right) \, \mathrm{d}\Delta = \pi \ln \frac{r_0}{r_1} \, .$$

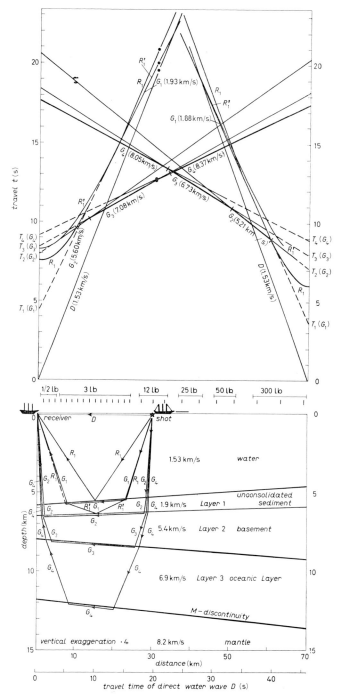

Fig. 2. – A deep sea, seismic refraction profile showing observed travel time curves with the corresponding ray paths, and their structural interpretation (from TALWANI, ref. [35]).

To save time derivation will be skipped (it can be found in any standard seismology book).

$\Delta$ is the angular distance from source to receiver (seismograph), $p$ is the slope of travel time $-\Delta$ curve or $\mathrm{d}t/\mathrm{d}\Delta$, $p_1 = r_1/\alpha_1$ is the value of $p$ at $\Delta_1$, corresponding to the ray which penetrates to $r_1$, $r_0$ is the Earth's radius. Given a $t$-$\Delta$ curve from observations of $P$ or $S$ waves from explosions, $\alpha(r)$ is found. Low-velocity zones are masked and difficult to handle without auxiliary methods. A recent development is the use of large arrays of seismographs built to detect nuclear explosions. These arrays yield precise values of $\mathrm{d}t/\mathrm{d}\Delta$. Since for any $\Delta_1$, $\mathrm{d}t/\mathrm{d}\Delta = r_1/\alpha_1$, this leads to a high-resolution method for obtaining $\alpha(r)$. Figure 5 shows the latest $\alpha(r)$ obtained by joint inversion of $t$ and $\mathrm{d}t/\mathrm{d}\Delta$ data, using arrays (ref. [25]).

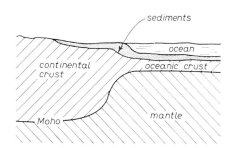

Fig. 3. – The most important contribution of seismic refraction profile was the discovery of crustal thickness variations between continents and oceans.

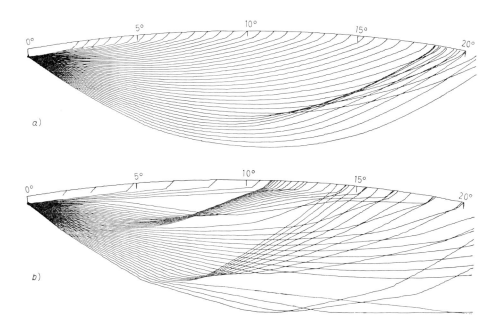

a)

b)

Fig. 4. – Trajectories or compressional waves in the upper mantle. Upper model contains one rapid increase in velocity. Lower model includes a low-velocity zone and two rapid transitions to higher velocity. (From ANDERSON, ref. [3]). a) JEFFREYS. b) NTS (East).

*Surface waves.* Body waves are the only elastic waves in an infinite medium without discontinuities. When surfaces, boundaries or other discontinuities occur, additional waves are possible. These have been called sur-

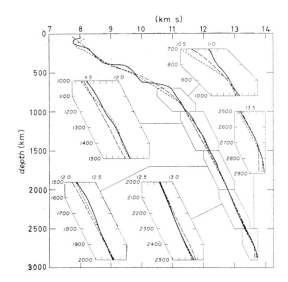

Fig. 5. – The classical Gutenberg and Jeffreys velocity distributions compared with Johnson's results based on data from large arrays (ref. [36]). – – – GUTENBERG. — — — JEFFREYS. ——— CIT 208.

face waves, interface waves, guided waves, etc. In the Earth the spherical free surface, the crustal layer and the velocity gradient in the mantle give rise to a family of surface waves (ref. [3, 26, 28]).

Solutions of eqs. (1) and (2) with appropriate boundary conditions yield characteristic equations which define the phase velocity in terms of the parameters of the wave guide. For example, the characteristic equation governing the propagation of Rayleigh waves in a liquid layer over a solid half-space is:

$$(6) \qquad \mathrm{tg}\left(kH\sqrt{c^2/\alpha_1^2-1}\right) = \frac{\varrho_2\beta_2^4\sqrt{c^2/\alpha_1^2-1}}{\varrho_1 c^4\sqrt{1-c^2/\alpha_2^2}}\cdot\{\ \},$$

where

$$(7) \qquad \{\ \} = 4\sqrt{1-c^2/\beta_2^2}\sqrt{1-c^2/\alpha_2^2}-(2-c^2/\beta_2^2)^2\,,$$

$c$ is phase velocity, $\alpha$, $\beta$, $\varrho$ are compressional and shear velocity and density respectively, subscripts 1 and 2 refer to the liquid layer and solid and $H$ is the layer thickness.  Group velocity is obtainable in the usual way from

$$(8) \qquad\qquad U = c + kH \, \mathrm{d}c/\mathrm{d}kH \, .$$

Equation (6) relates $c$ to dimensionless wave number $kH$ for the possible solutions with elastic constants $\alpha$, $\beta$ and $\varrho$ as parameters.  Solutions are shown graphically in Fig. 6.

Thus the solutions are dispersive, long waves travel with velocity $C_R$(*), short waves with velocity $\alpha_1$, in the first mode as would be expected.  Infinite numbers of modes are possible—which ones are excited depends on the position and spectrum of the source.  By obtaining dispersion curves from seismic records one can say something about $H$, $\varrho_1$, $\varrho_2$, $\alpha_1$, $\beta_2$.  If we know in advance that we are dealing with a single liquid layer over half-space, $\alpha_1, \beta_1, \beta_2$ can be uniquely determined.  $H, \varrho_1, \varrho_2$ can be inferred with some reasonable assumptions.  This is the principle of surface wave analysis to reveal internal Earth properties.  In practice it is more complicated because many more layers are involved.

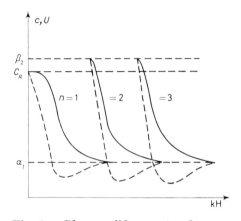

Fig. 6. – Phase (solid curves) and group (dashed curves) velocities for 3 modes of the system liquid layer over a half-space. $C_R$ is the root of eq. (7).

Figure 7 shows actual records of dispersed Rayleigh waves (from an earthquake in the S.W. Pacific ocean) propagating across an array of stations in the western U.S.  The dashed line shows how phase velocity is found for a particular period.  Group velocity is found from the travel time of a wave group of a fixed period.  Note that the dispersion originates in this case primarily from the effect of the ocean.

Despite the simplicity of eqs. (6) and (8), they can be used to interpret these records and conclude that the sea floor unlike the continents lacks a significant thickness of granite.  The more complex situation of a multilayered crust overlying a vertically heterogeneous mantle in a spherical Earth is handled on computer.  Programs yield phase and group velocity curves for realistic models for comparison with experimental values.

Experimentally determined Rayleigh wave dispersion curves for the amazingly large range $(1 \div 2000)$ s are shown in Fig. 8 (see ref. [28]).  The gray

(*) $C_R$ is the velocity of Rayleigh waves in the half-space.

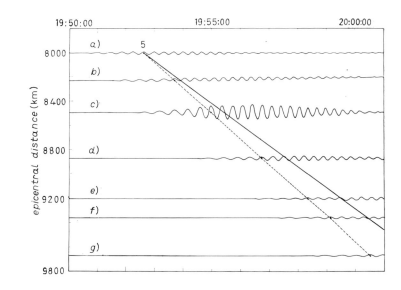

Fig. 7. – Dispersed Rayleigh waves propagating across an array of stations in North America. Records like these yield phase and group velocity as a function of period. —— Group velocity. — — — Phase velocity. *a*) Pasadena. *b*) Reno. *c*) Eureka. *d*) Salt Lake City. *e*) Bozeman. *f*) Laramie. *g*) Rapid City.

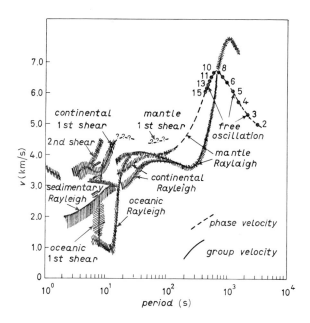

Fig. 8. – Experimentally determined Rayleigh wave phase and group velocity curves for the Earth in the period range (1÷3230) s (after OLIVER, ref. [28]).

area shows scatter, probably due to lateral variations in crust and mantle properties.

Thus far we have been dealing with wave motion in which particles vibrate in the vertical plane of propagation. Another class of waves is characterized by particle motion in the horizontal plane and normal to the direction of propagation. These are Love waves and the corresponding dispersion curves are in Fig. 9.

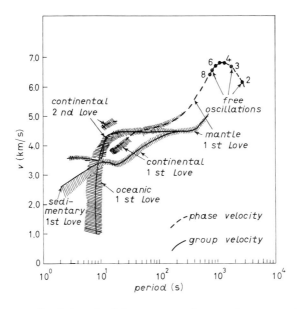

Fig. 9. – Experimentally determined Love wave phase and group velocity curves for the Earth in the period range $(1 \div 3230)$ s (after OLIVER, ref. [28]).

## 3. – Free oscillation of the Earth.

One of the most remarkable experiments of the past decade was the detection and identification of the Earth's free oscillations. In essence these waves are equivalent to the global standing wave pattern formed by interfering Rayleigh and Love waves. The Rayleigh waves correspond to the spheroidal oscillations and the Love waves correspond to the torsional oscillations. The periods of the oscillations depend on $\alpha(r)$, $\beta(r)$, $\varrho(r)$ and gravity. Both $S$ and $T$ vibrations have double infinities of modes—i.e., there is an infinite number of modes, each mode has an infinite number of overtones. Which ones are excited depends upon the position, mechanism and spectrum of the source.

For the simplifying case of azimuthal symmetry the displacement $V$ cor-

F. PRESS

responding to the $n$th mode has the form

(9) $$V_n = \sum_{m=0}^{\infty} F(r, {}_m\omega_n) P_n(\cos\theta) \exp\left[i {}_m\omega_n t\right].$$

The standing wave pattern on the surface of the earth is described by the surface zonal harmonic $P_n(\cos\theta)$. Thus $n = 2$ is the football mode of spheroidal oscillations and $n = 0$ is the radial mode.

The term $F(r, {}_m\omega_n)$ describes the vertical variation pattern, the overtone index $m$ indicating the numbers of internal nodal surfaces. The eigenfrequencies have been identified for 200 fundamental ($n = 0$) modes and for some 20 overtones. Propagating and standing waves are related by $n + \frac{1}{2} = r_0 {}_m\omega_n/{}_m c_n$. Thus a knowledge of any two of $n, c, \omega$ can be used to obtain the third factor. In practice Earth structure is studied with free oscillations for $n < 25$ and with propagating waves for $n \geqslant 25$ as a matter of experimental convenience.

An observation of free oscillations is made by performing a spectral analysis

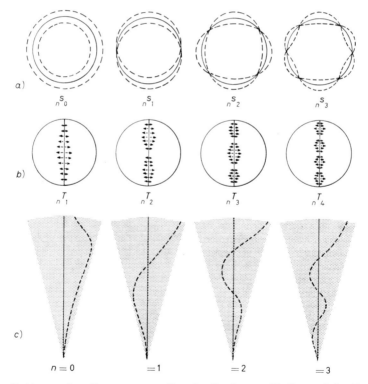

Fig. 10. – Patterns of motion corresponding to the free oscillations of the Earth. Each surface pattern designated by the subscript to the right is associated with an infinite number of overtones with depth patterns indicated by the subscript $n$. a) Surface patterns (spheroidal). b) Surface patterns (torsional). c) Depth patterns.

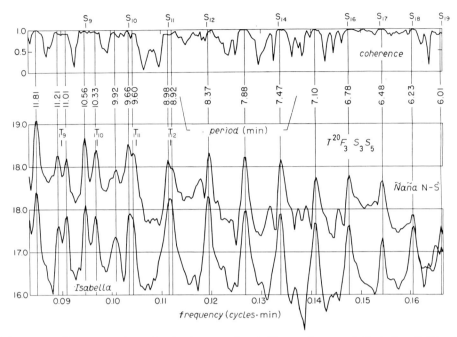

Fig. 11. – Power spectra of seismic records at Nana, Peru and Isabella, California revealing the spectral peaks corresponding to the free oscillations excited by the great Chilean earthquake of 1960.

on records of long period seismographs or gravimeters following a large earthquake. The records are digitized automatically and analysis proceeds on high speed digital computers. Eigenfrequencies can be determined with a precision of about 1 part in 500.

Figure 10 shows surface and depth patterns corresponding to several modes. Figure 11 shows power spectra whose peaks are used to determine eigenfrequencies.

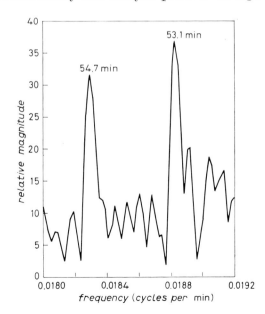

Fig. 12. – High resolution Fourier analysis reveals the splitting of the free oscillation mode $_0S_2$ due to rotation of the Earth which affects the symmetry between waves traveling in opposite directions.

The earth is not spherical, it rotates, and is laterally heterogeneous. All of these effects « split » the spectral peaks, making it difficult to obtain precisely the central frequencies which are needed to infer the properties of the interior.

An example of the mode spitting due to rotation is shown in Fig. 12.

Since the eigenfrequencies depend on the elastic properties of the interior they are a principal tool to recover these properties. The procedures used will be discussed in an Appendix to this chapter where travel time, $dt/d\Delta$, surface wave data and eigenfrequencies will be combined to make statements about the interior. The problem of « nonuniqueness » of solution will be discussed in the Appendix and in later chapters.

## 4. – Contributions from geological and laboratory studies.

The goal of geophysics is not achieved by inferring the distributions $\alpha(r)$, $\beta(r)$, $\varrho(r)$ for these parameters are only important in so far as they enable us to obtain information about composition, physical state and dynamic processes. What is needed is a connection between the parameters measured in « outdoor » geophysical experiments and the Earth's internal constitution and state. This connection is supplied in the laboratory. The rocks which make up the Earth's crust are accessible to geologists. Samples of the upper mantle are recovered from volcanoes and diatremes although it is debatable whether or not these are altered in the process of ejection to the surface. Meteorites and Moon rocks provide clues about the interior of other planets. These materials are subjected to several types of laboratory studies.

1) Artificial rocks are created in the laboratory, duplicating the mineral assemblages in natural rocks. Experiments can be carried out at temperatures at high as 1500 °C and pressures of $10^5$ atmospheres—equivalent to depths of about 400 km in the Earth. In this way mineral assemblages can be calibrated as thermometers and pressure gauges with which to determine the depth of origin of rocks.

2) Liquid-solid and solid-solid phase changes are studied as a function of $P$ and $T$. Since shear velocity is very sensitive to the presence of even a small amount of melt, sudden decreases in $\beta(r)$ can be related to the solidus within the Earth. Melting points of common rocks like granite, basalt, peridotite, pyrolite (*), dunite and minerals like olivine, and abundant elements like iron are studied in the laboratory.

_____

(*) Pyrolite is a rock similar to peridotite, whose composition is such that its first melt is basalt.

Solid-solid phase changes are also important in the Earth; graphite-diamond is a famous transition which occurs in the Earth—and is useful as a depth indicator. Diamond is so rare that it is not a factor in the bulk properties of the

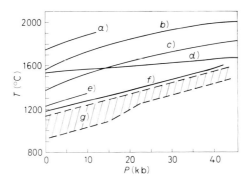

Fig. 13. – Melting temperatures of rocks and minerals are used to infer physical state in the Earth's interior. It is probable that the mantle between 100 and 200 km is partially molten, the melt being basalt (ref. [29]). a) Forsterite. b) Enstatite. c) Diopside. d) Iron. e) Fayalite. f) Pyrolite. g) Basalt.

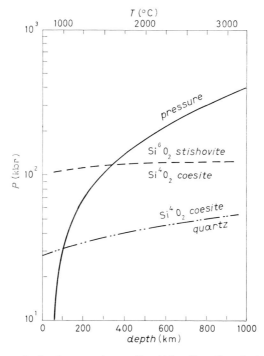

Fig. 14. – Phase boundaries for quartz-coesite-stishovite. Quartz transforms to coesite at 100 km and the transition to stishovite occurs near 350 km. (Ref. [30].)

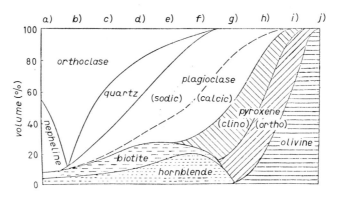

Fig. 15. – Approximate mineral compositions of igneous rocks which make up the crust and mantle (ref. [31]). *a*) Nephline syenite (phonalite). *b*) Syenite (trachyte). *c*) Granite (rhyolite). *d*) Granodiorite (rhyodacite). *e*) Quartz diorite (dacite); *f*) Diorite (andesite). *g*) Gabbro (diabase) (basalt). *h*) Olivine gabbro (olivine diabase) (olivine basalt). *i*) Peridotite. *j*) Dunite.

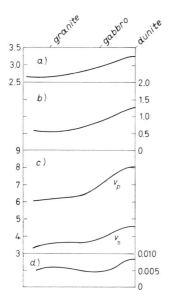

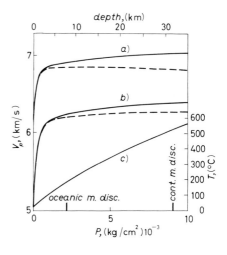

Fig. 16. – Physical properties of igneous rocks of Fig. 15, using same horizontal order. This enables us to associate composition at depth with physical properties determined from surface observations (ref. [31]). *a*) Density. *b*) Incompressibility mbar. *c*) Velocity km/s. *d*) Thermal conductivity 200°.

Fig. 17. – Velocity increases with pressure, hence depth in the Earth. Temperature acts in the opposite direction. The heavy curves show the effects of pressure and temperature for the two rock types which are the major constituents of the crust (see ref. [32]). *a*) Gabbro. *b*) Granite. *c*) Temperature.

Earth's mantle. It is a major accomplishment of recent years that a phase change of olivine (*) to the spinel structure was achieved in the laboratory and the results correlated almost perfectly with the seismic transition near 400 km (see Fig. 5), delineated by $dt/d\Delta$-array observations.

3) Elastic velocities and density as functions of $P$ and $T$. These laboratory results have veen carried to 1000 °C and 20 000 atmospheres—equivalent to about 60 km in the Earth.

4) Shock waves studies where atom-bomb implosion technology is used to obtain instantaneous pressure of $10^6$ atmospheres at several thousand degrees C. Density and shock velocity can be recovered, but since the process is neither isothermal nor adiabatic the reduction of $\varrho$ and $\varphi$ to adiabats of a given temperature for earth comparisons is a little uncertain. The results are good enough at present to make some gross chemical conclusions, as we shall see later.

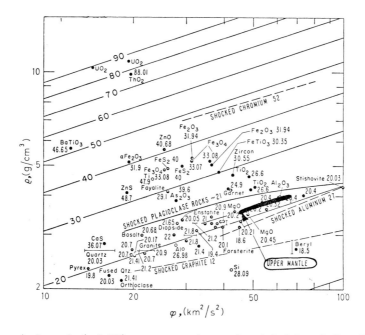

Fig. 18. – Anderson's (ref. [3]) summary of experimental data relating the seismic factor and density with mean atomic weight as a parameter. Superimposed is a band of upper mantle earth models derived in the Appendix. The range of possible mantle constituents can be inferred in this manner.

---

(*) Olivine, a solid solution $(Mg, Fe)_2SiO_4$ is perhaps the most abundant mineral in the mantle.

Figure 13 (ref. [29]) shows some melting point relations for common rocks and minerals. The presence of small amounts of water would lower the solidus below that shown—a factor which may be important in the Earth.

Figure 14 (ref. [30]) shows an example of a solid-solid phase transition: quartz-coesite-Stishovite. Coesite and stishovite have been found at the sites of nuclear explosions and meteor creaters where the transition was generated by shock pressures associated with these events.

Figures 15-17 (ref. [31]) review the mineral compositions of igneous rocks and show how velocity and density vary with composition and temperature.

Figures 18 and 19 (ref. [3]) are a summary of shock wave data for numerous elements, compounds, minerals and rocks. The seismic-shock wave factor $\varphi$

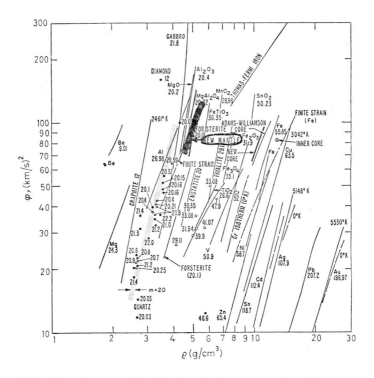

Fig. 19. – Same as Fig. 18, for the lower mantle and core.

is plotted against $\varrho$ with mean atomic weight as a parameter. When seismic results are superimposed on a diagram like this, as will be done later, the conclusion is inescapable that mean atomic weight $\overline{m}$ for the Earth's mantle lies in the range $20 < \overline{m} < 22$ and that the Earth's core is mostly iron alloyed with a « lightening » element.

## 5. – Some major conclusions on Earth structure.

We are now in a position to review some of the important results which have been achieved using the preceding methods. An Appendix dealing with the practical aspects of inversion of geophysical data to obtain structure is at the end of this chapter.

1) *Crustal structure*. Crustal thickness, velocity and density vary in the complex way shown in Fig. 20. The continental crust, with thickness about 35 km is composed of granites and basalts—a light weight $(\varrho \sim (2.7 \div 2.9))$ chemical differentiate of the mantle with major minerals—feldspars, quartz, pyroxene, mica, hornblende (mostly $SiO_2$, $Al_2O_3$, CaO, MgO, FeO) (see Fig. 15).

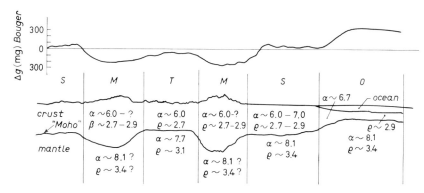

Fig. 20. – Regional variations in crustal and upper mantle structure revealed by seismic refraction and gravity observations. $S$) Continental shield. $M$) Mountain. $T$) Tectonic. $O$) Oceanic (abyssal plains).

The ocean crust is basaltic and only about 5 km thick. Topographic loads such as continents with respect to oceans or mountains with respect to shields are compensated by crustal thickening, *i.e.* by a light crustal «root» projecting as much as $(50 \div 60)$ km into the more dense mantle. In tectonic belts with moderate topography compensation can be achieved with a shallow, slightly less dense mantle.

The $M$-discontinuity is a chemical boundary which separates the crust from the more dense parent mantle which is predominantly olivine and pyroxene ($SiO_2$, MgO, FeO).

These results stem primarily from seismic refraction, gravity and geological observations.

2) Seismic surface and body wave analyses reveal that the upper mantle is zoned as in Fig. 21. There is of course more uncertainty than is implied by

the diagram but I believe it is fairly close to reality because of the interplay of geologic, seismic and laboratory data which are jointly interpreted in reaching these conclusions. For example, lavas are known to come from depths of about 70 km; the latest laboratory results on melting in the presence of water have lowered the solidus so that it is intersected by almost every conceivable

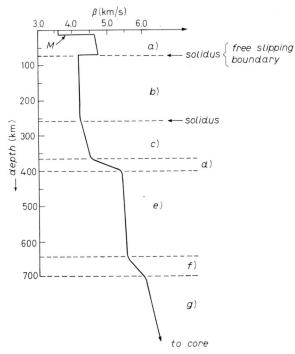

Fig. 21. – Structure of the Earth's mantle inferred from recent, seismological studies. *a*) Lithosphere-plate solid, strong, high velocity, high-$Q$ laterally inhomogeneous. *b*) Asthenosphere $(1 \div 10)\%$ partially molten, weak. Low velocity, low $Q$, source or basaltic lava, laterally inhomogeneous. *c*) Solid, high $Q$, self compression. *d*) Phase change olivine-spinel. *e*) Solid, high-$Q$, self compression. *f*) Phase change, possible increase in Fe/Mg, breaking down to FeO, MgO, $SiO_2$ and $Al_2O_3$ 2. *g*) Solid, high-$Q$, possibly superadiabatic or decreasing Fe/Mg, small lateral inhomogeneities.

geotherm near 70 km; every model for $\beta(r)$ under oceans requires a low-velocity zone beginning near 70 km; the plate tectonics hypothesis requires free slipping at the base of the lithosphere; seismic data show high $Q$ above about 70 km and low $Q$ below. These various sources of data and information converge to a self-consistent explanation indicated for the upper mantle shown in Fig. 21.

3) In a similar manner the abrupt velocity increase at $(350 \div 400)$ km shows up clearly on the $dt/d\varDelta$ data and even in the absence of these data would

have been predicted from the geological abundance of olivine and the laboratory conversion of olivine to spinel at pressures corresponding to this depth.

4) The mantle from about 1000 km to the core shows no major discontinuities or rapid changes—implying a zone which is to first order chemically
homogeneous, and nearly adiabatic without major phase changes. The data
are also consistent with a slight superadiabatic gradient or decreasing Fe/Mg
ratio. From Fig. 18 and 19 where earth models are compared with shock data,
we conclude that $20 < \overline{m} < 22$ is about as good a description possible with
present technology. $\overline{m}$ is primarily an indicator of the Fe/Mg ratio, which probably lies in the range $1/10 < Fe/Mg < 2/10$ for the mantle. It may be surprising that gross chemistry can be obtained from seismological data. Actually
the range of possibilities is too broad as it includes most geochemical hypotheses
and cannot therefore distinguish between them.

5) The fluid core densities obtained from free oscillation data (Fig. 22)
are shown together with shock-derived densities for Fe, Ni and Fe $+$ 19.8 Si.
Iron is the only element sufficiently abundant to supply most of the mass of
the core. The Earth models imply den-
sities somewhat less than Fe suggesting
the presence of a lightening element.
On geochemical grounds the candidates
are Si, S, O. If Si is the agent then
about $(10 \div 15)\%$ Si content is implied
by the range of models in Fig. 22.

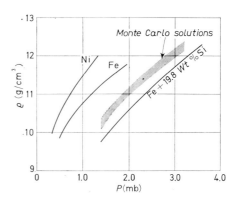

Fig. 22. – Core densities derived from the
inversion procedures in the Appendix compared with shock densities of constituents
of the core.

## 6. – Tectonics.

Up to this point we have been primarily concerned with Earth structure.
In this Section we introduce the subject of tectonics. The two topics are not
unrelated. Partial melting gives rise to buoyant fluids, a strong lithosphere
is capable of rigid-body motion over the asthenosphere—these are the in-
gredients of a complex convection system involving solid and fluid flow—with
energy release in the form of earthquakes, opening oceans, drifting continents,
orogeny (mountain making), volcanism etc. The motivation of global tectonics
is therefore to understand the dynamic processes which shape the Earth's
surface and which have led to the environmental conditions which have made
life possible and hazardous at the same time.

6.1 *Kinematical tectonics.* – In the history of the geophysical sciences the decade of the sixties will be memorable because of the hypothesis of plate tectonics which has been verified in so many waves as to qualify as a confirmed theory. Briefly the hypothesis proposes to divide the lithosphere into the order of 10 large, rigid blocks or plates in relative motion with respect to each other. Earthquakes occur at the boundaries between plates. The plates spread apart at ocean ridges where new material is added. The plates slide past one another along strike-slip transform faults. Where plates converge at island arcs, one plate underthrusts the other and sinks back into the upper mantle where it heats up and finally melts (see ref. [32-34]).

6.1.1. S e i s m i c  e v i d e n c e  f o r  p l a t e  t e c t o n i c s. With data from hundreds of modern seismograph stations processed by large computers, earthquake hypocenters can be located very precisely. Figure 23 is a plot of earthquake locations for the year 1966 by the U.S. Coast and Geodetic Survey. The epicenters are confined to narrow continuous belts that bound large, aseismic blocks. The western S. Atlantic and South America form one block bounded on the east by the earthquakes of the mid-atlantic ridge and on the

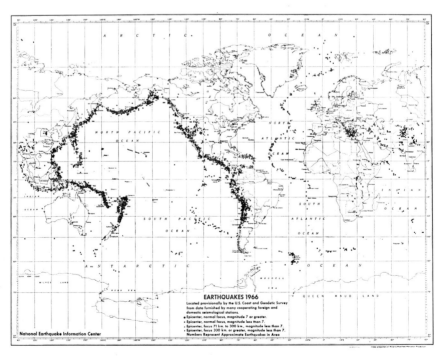

Fig. 23. – The precision location of earthquakes has made possible the delineation of the rigid plates whose relative motions are responsible for the primary tectonic features of the Earth.

west by the earthquakes of the deep sea trench at the Pacific border. Where the plates diverge the hypocenters are shallow. Where the South American plate overides the Pacific plate, the hypocenters lie on an oblique plane extending from the surface in the trench to depths of about 700 km under the Andes.

Additional seismic evidence for plate tectonics is derived from source mechanism studies. An earthquake is a sudden slip along a fault and the radiation pattern of seismic waves reflects the mechanism, *i.e.* the slip direction, and the fault plane orientation. An example is shown in Fig. 24 (ref. [32]) where typical earthquake mechanisms are shown for a convergent zone. The $+$, $-$, symbols represent initial $P$ wave motions towards or away from the source. Waves leaving the source in the different quadrants arrive at seismograph stations with this polarity preserved. The station data are used to reconstruct the fault and slip directions. Note that there are two possible source and slip directions —one of which can often be selected on the basis of auxiliary

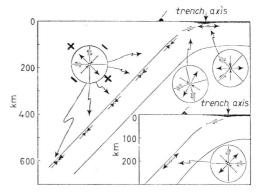

Fig. 24. – Earthquake mechanisms found from the radiation of seismic waves from earthquakes in a convergent zone. Axes of compression and tension are indicated by arrows in the lithospheric slab. The circular blow-ups show the radiation pattern and the sense of motion on the two possible slip planes (ref. [32]).

data such as the definition of the fault plane by geologic evidence or the location of aftershocks. In any case the principal stresses shown by converging or diverging arrows are recoverable.

On a world-wide basis the midocean ridge divergence zones show tensile stresses. Transform faults show strike-slips (horizontal slips consistent with relative motion between two plates slipping past one another) and convergent

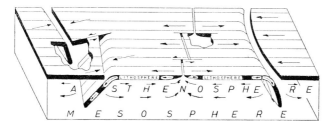

Fig. 25. – Block diagram depicting lithosphere divergence and convergence zones, transform faults and return circulation in the asthenosphere (ref. [32]).

zones show the pattern in Fig. 24. All of the mechanisms are consistent with plate tectonics as can be seen in the block diagram of Fig. 25 and the summary map of slip vectors (derived from source mechanism studies) shown in Fig. 26, taken from ref. [32].

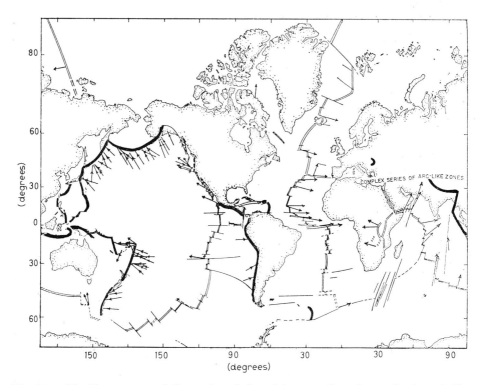

Fig. 26. – World summary of slip vectors deduced from earthquake mechanism studies. Double lines represent divergence zones, bold solid lines convergent zones and thin single lines depict transform faults (ref. [32]).

The sinking lithosphere has been verified in a more direct way. Since it is a plate with higher velocity and higher $Q$ than the asthenosphere it can be expected to conduct seismic waves more rapidly and efficiently. Both of these effects have been verified.

6˙1.2. Evidence from fossil magnetism. Actually this was the first geophysical evidence presented in support of the hypothesis. It is elegant in its simplicity. The lithospheric plate is formed in the diverging zones from rising magma which cools, solidifies and is translated bilaterally away from the ridge with the moving plates. In cooling through the Curie point the rock becomes magnetized in the direction of the Earth's magnetic field. As will be discussed in later lectures the Earth's field reverses polarity at intervals of about $(10^5 \div 10^7)$ years and this pattern of reversals is imprinted and preserved in the rock as shown in Fig. 27. Measurements of the magnetic field using ships or air planes traversing lithospheric plates normal to a zone of divergence reveal a characteristic pattern of positive and negative fluctuations which are used to infer the magnetic imprinting in the rocks below. The history of

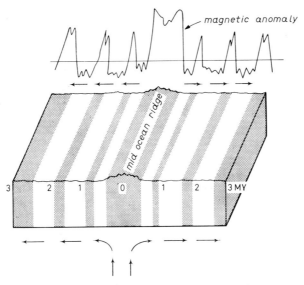

Fig. 27. – Schematic view of ocean floor spreading as evidence by sequences of normally and reversely magnetized sections of the oceanic crust. The direction of movement is indicated by arrows and the time scale in millions of years. The magnetic anomaly corresponding to the magnetized rocks is also shown. ☐ Normal polarity. ☐ Reversed polarity.

reversals of the Earth's field has been unraveled for the past several million years, primarily from analyses of successive lava flows which have been dated by radiochemical methods. The magnetic fluctuations observed in surveys over spreading plates can therefore be used to obtain rates of spreading. Values in the range $(1 \div 5)$ cm/yr have been obtained. Figure 27 shows the characteristic magnetic anomaly patterns for the Pacific-Antarctic ridge and for the mid-Atlantic ridge. The expected symmetry of the magnetic fluctuations with respect to the ridge axis is evident in the Figure.

6'1.3. Evidence from deep-drilling. Perhaps the most conclusive evidence in favor of the plate Tectonics hypothesis comes from deep drilling on the sea floor. A special drilling ship has been constructructed which can drill at mid-ocean depths through the entire sedimentary column to the basaltic basement. A profile of drilling was carried out last year starting at the mid-Atlantic ridge and crossing the ocean. The ages of the cores were obtained by palontological and geochemical methods. The sequence of young sediments progressing to older sediments with increasing distance from the ridge was exactly that predicted by the hypothesis and the rate of spreading matched to about 1% that calculated by magnetic methods (ref. [34]).

6'1.4. Other sources of evidence. All of the evidence, albeit circumstantial, connected with the old theory of continental drift is pertinent. Perhaps the most famous is the « jig-saw puzzle » fit of the continents before the Atlantic Ocean opened up some $10^8$ years ago.

Heat flow from the Earth's interior can be obtained from measurements of temperature gradients and thermal conductivity near the surface. Thousands of observations have been made. The expected high heat flow is found over the diverging zone where the rising magma delivers heat to the surface much more efficiently than the conductive flow which is the primary mechanism elsewhere.

6'2. *Dynamical tectonics*. – The mobility of the lithosphere is now well understood, at least in a descriptive sense. The dynamics of plate tectonics is now the major unresolved problem. What follows are some qualitative elements of my own hypotheses (ref. [23]). I will leave it to the other lecturers to present their views.

The asthenosphere is partially molten, weak and flow readily occurs. The first melt of a periodotic mantle are its heaviest components—but in the fluid state the magma rises buoyantly and turns over at the base of the lithosphere. The drag produces a fracture which enables the magma to percolate through the lithosphere to the surface. Upon cooling the magma attaches itself to the lithosphere and becomes more dense than the parent material. Thus the lithosphere is gravitationally unstable with respect to the asthenosphere. The gravitational instability together with the drag near the diverging zone provide the driving push-pull system of stresses. This simple view will ultimately be replaced by an analysis of the complex convection pattern of a single system of which the lithosphere and asthenosphere are components.

APPENDIX

## A practical example of inversion of geophysical data.

What follows is one example of the inversion of geophysical data, drawn almost entirely from ref. [21]. The idea is to use the available geophysical data such as the mass and moment of inertia of the Earth, the travel time of seismic body waves, the phase velocity of surface waves, and the eigen-frequencies of the Earth in a procedure which yields the vertical variation of compressional velocity, shear velocity and density in the Earth. A final step is to interpret these distributions in terms of composition and physical state. In a formal sense the problem is exceedingly difficult because unique distributions cannot be found. This follows from the incompleteness of the data set. Another difficulty is presented by the errors in the data and their uncertain distribution. Discussions of the several procedures currently practiced in the inversion of geophysical data will be found in later lectures by Professor KNOPOFF and Professor KEILIS-BOROK. In this Appendix we outline the Monte Carlo approach which offers the pedagogical advantage of providing a band of possible solutions which fit the data. The width of the band of solutions shows the degree to which the structure is constrained by the data in some parts of the Earth and is essentially unconstrained in other parts. Uncertainties in the data are taken into account in a natural way and unconscious biasses are removed.

## Method.

The Monte Carlo procedure is described by the flow diagram in Fig. 28. An algorithm is used to generate random values of compressional velocity, shear velocity, and density as a function of depth, defining a specific Earth model. Actually the random selection is required to fall within upper and lower bounds which are set with wide enough separation to ensure that the real Earth falls within the bounds. The random models are tested against the travel times of compressional and shear waves and the mass and moment of inertia of the Earth. When the model is found which satisfies these constraints a final test is made against the eigenfrequencies of the Earth. If the model fails any test the program loops back to an earlier stage where a new model is generated. Monte Carlo methods are most powerful when a sufficiently large number of models are examined so that the retained solutions are representative of the class of successful models which fit the data. This requires an extremely fast program. The high speed is achieved in several ways. Eigenfrequencies are computed by variational parameters rather than by numerical integration of the differential equations. Approximately 25 000 variational parameters are entered into the computer from disc memory during each run of the program. This procedure makes possible a rapid and accurate computation of eigenfrequencies by table look up and is the essential feature of the method. Without it Monte Carlo techniques would be prohibitive in computer costs because of their inherent inefficiency. The variational parameters used are:

$$\partial T/\partial\varrho_i, \qquad \partial T/\partial\beta_i, \quad \text{and} \quad \partial T/\partial\alpha_i$$

and the relationship

$$\Delta T = \sum_i \left( \frac{\partial T}{\partial\varrho_i}\,\Delta\varrho_i + \frac{\partial T}{\partial\beta_i}\,\Delta\beta_i + \frac{\partial T}{\partial\alpha_i}\,\Delta\alpha_i \right)$$

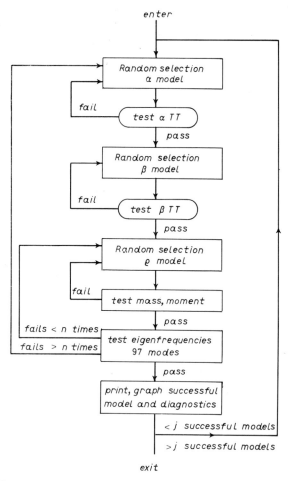

Fig. 28. – Flow diagram of Monte Carlo inversion program.

to compute the change in period $T$ for a particular mode due to perturbations $\Delta\varrho_i$, $\Delta\beta_i$, and $\Delta\alpha_i$ in density, shear velocity, and compressional velocity respectively, of the $i$-th layer. A more complete description of the constraints placed on the models can be found in ref. [22]. Reference [15] is an early description of the use of Monte Carlo methods in geophysics.

## Results.

Preliminary to finding models which fit the data, a test was made to see that randomly selected models uniformly filled the space between the permissible bounds. A run in which 25 models were randomly selected but not tested against the data is shown in Fig. 29. No bias is apparent in the selection procedure.

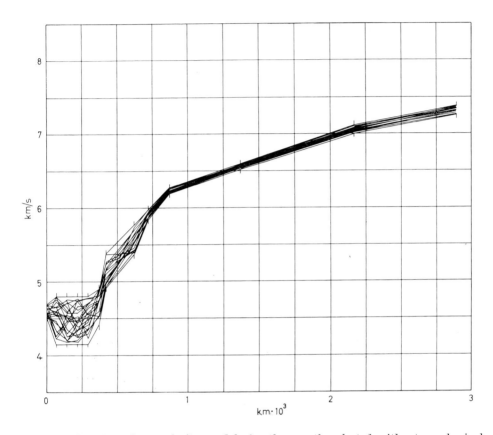

Fig. 29. – Random shear velocity models for the mantle selected without geophysical constraints to test for bias in the selection procedure.

The results of a search for models which fit the data are presented as follows:
Figures 30-31. Density models for a mantle topped by oceanic structure.
Figure 22. Density models for the core.
In each group of models the upper and lower bounds within which models were selected for testing are shown by heavy lines.

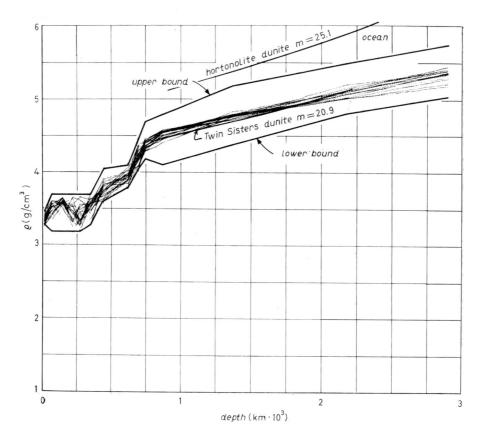

Fig. 30. – Successful density models for the mantle compared with predicted densities for dunites with varying mean atomic weight.

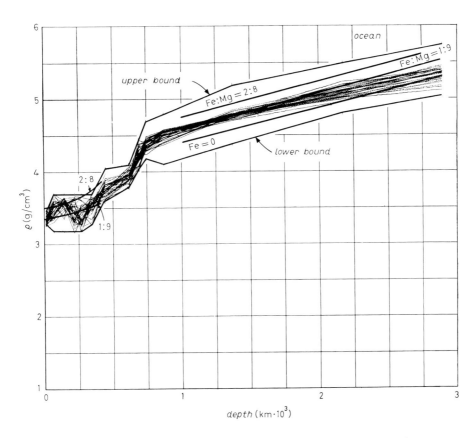

Fig. 31. – Successful density models for the mantle compared with olivine-pyroxene mantles with varying Fe/Mg ratios.

## Interpretation of results.

Over most of the range below 800 km the density solutions fall within a band of width 0.25 g/cm³. Superimposed on the density models in Fig. 30 are the expected density, based on shock-wave data, for Twin Sisters dunite with mean atomic weight $m = 20.9$ and hortonolite dunite with $m = 25.1$ (ref. [2]). The lower mantle solutions are consistent with constant composition ($m \sim 21$) or a changing composition in the direction of a reduction in $m$, which is primarily an indicator of the Fe/Mg ratio. This can also be seen in Fig. 31 where the same solutions are presented with Fujisawa's (ref. [10]) estimated densities for varying Fe/Mg values. In computing density Fujisawa made the reasonable assumption that the Fe/Mg ratio in olivine and pyroxene is approximately the same, and that both species break down into a mixture of oxides MgO, FeO and $SiO_2$ under lower mantle conditions. Figure 31 implies an Fe/Mg ratio of $\frac{1}{9}$ for the mantle below 800 km or a gradual reduction in this ratio from a beginning value of $\frac{1}{9}$ at about 800 km. Although a superadiabatic temperature gradient would show a density variation similar to that for a reduction in Fe/Mg this possibility is minimized since the corresponding increase in the gradients of compressional and shear velocities are not present.

Similar results are obtained by a comparison of density $\varrho$ and bulk sound velocity $c = \sqrt{\varphi}$, as obtained from the Monte Carlo solutions with values for dunite of varying Fe/Mg content, as calculated from shock-wave data. Unfortunately the reduction of the shock-wave data by two different groups leads to somewhat different $c$-$\varrho$ plots. Wang's (ref. [24]) calculation for Twin Sisters dunite is a straight line connecting the two crosses in Fig. 32. The slope of this line differs from that found by AHRENS, ANDERSON and RINGWOOD (ref. [1]), using the same shock wave data. The difference persists for the more iron-enriched materials. Despite these differences the results indicate a lower mantle with $m \sim (21 \div 22)$, either chemically homogeneous or decreasing in iron content with depth (*).

CLARK and RINGWOOD (ref. [8]) recognized that a $c$-$\varrho$ comparison of Earth models with laboratory data offers the advantage of minimizing the effect of temperature to first order, since the proportional changes in $c$ and $\varrho$ are the same for a given temperature variation.

The offset in the band of solutions on both sides of the depth range $(620 \div 870)$ km in Fig. 32 suggests an increase in iron content with depth through the lower part of the transition zone corresponding to $\Delta m \sim 1$. The region $(350 \div 400)$ km where olivine transforms to the spinel structure shows no offset, implying unchanging composition.

The shear velocities in the lower mantle fill the permissible range indicating that the travel time and $dt/d\Delta$ data which are used to establish the bounds are more constraining than the eigenperiod data.

The density and elastic velocities in the upper mantle are constrained by seismic refraction, surface wave and eigenperiod data. The refraction experiments place bounds on the initial velocity (hence density) just below the

---

(*) *Note added in proofs.* – Recent work indicates that Wang's reduction should be used.

$M$-discontinuity. The surface-wave data provide whatever resolving power is available for upper mantle structures and the eigenperiod data make the upper mantle solutions consistent with the whole Earth. For example, absolute densities for the upper mantle cannot be assigned without fitting with high

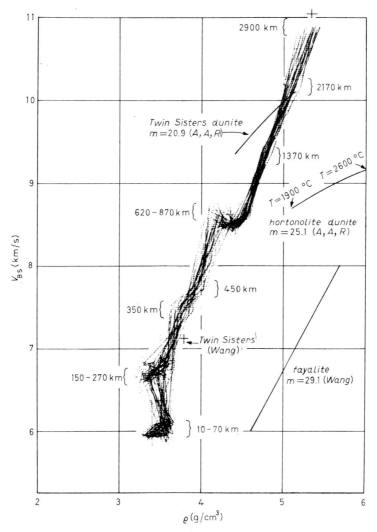

Fig. 32. – Plot of bulk sound velocity $(\varphi)^{\frac{1}{2}}$ vs. density from successful solutions and comparison with shock-wave-derived curves for ultrabasic rocks.

precision the graver eigenmodes. Another example—the resolution of the low-velocity zone of the suboceanic mantle deteriorates if the initial shear velocity at the $M$-discontinuity is allowed to take values below that indicated by refraction measurements.

The upper mantle density solutions in Fig. 30 and 31 are all characterized by positive initial density gradients and by an absence of solutions with $\varrho \sim 3.3$ g/cm³ between 70 and 150 km. The mean density over the depth range $(70 \div 170)$ km exceeds 3.4, implying material closer to eclogite in composition than pyrolite or peridotite. The scatter of density solutions below 150 km shows a lack of resolving power at these depths for the oceanic and shield models. A reversal of density is indicated in some models whereas others increase monotonically, implying an inability to resolve a low-density zone with these data.

The indication of a high density $(\varrho > 3.4$ g/cm³) cap to the mantle is an important result and deserves further verification. ANDERSON (ref. [2]) reports an upper mantle density of at least, 3.5 g/cm³. A numerical search for lower density models was attempted in the selection procedure by placing the upper bound at $(3.4 \pm 0.07)$ g/cm³ for depths above 140 km. No models with $\varrho < 3.4$ g/cm³ were found at depths below 50 km, the successful models all falling above this value. The solutions all show higher densities than the pyrolite models despite their origin in a search for lower density models. With the three point parameterization used in this search, there are about 200 significantly different density models for the region above 140 km. Since several hundred thousand density models were generated in this numerical test we feel that the density space was adequately examined in the search for lower density models.

Every model of shear velocity distribution in the mantle beneath oceans and continents requires a low velocity channel. The suboceanic channel tends toward lower velocities than occurs under shields. If the thickness of the lithosphere is defined arbitrarily by the depth at which the shear velocity drops below 4.5 km/s, then the sub-shield lithosphere is twice as thick as the sub-oceanic lithosphere.

The density distribution in the fluid, outer core falls within a surprisingly narrow band of about 0.3 g/cm³. Apparently the combination of eigenperiod data, mass, moment of inertia and the requirement of adiabaticity constrain the densities in the outer core. The wide divergence of solutions for the inner core imply that this region is uncontrolled by the data presently available.

REFERENCES

[1] T. J. AHRENS, D. L. ANDERSON and A. E. RINGWOOD: *Rev. Geophys.*, **7**, 667(1969).
[2] D. L. ANDERSON and T. JORDAN: *Phys. Earth Planet. Int.*, **3**, 23 (1970).
[3] D. L. ANDERSON: *The Earth's Mantle*, edited by T. F. GASKELL (London and New York, 1967).
[4] D. L. ANDERSON: *Petrology of the mantle*, in press (1970).
[5] G. E. BACKUS and F. GILBERT: *Geophys. Journ.*, **16**, 169 (1968).
[6] F. BIRCH: *Journ. Geophys. Res.*, **57**, 227 (1952).
[7] F. BIRCH: *Phys. Earth Planet. Int.*, **3**, 178 (1970).
[8] S. P. CLARK and A. E. RINGWOOD: *Rev. Geophys.*, **2**, 35 (1964).
[9] A. M. DZIEWONSKI: *Bull. Seismol. Soc. Amer.*, **60**, 741 (1970).

[10]  H. FUJISAWA: *Journ. Geophys. Res.*, **73**, 3281 (1968).

[11]  A. L. HALES and J. L. ROBERTS: *Shear velocities in the lower mantle and the radius of the core*, in press (1970).

[12]  K. ITO and G. C. KENNEDY: *The fine structure of the basalt-eclogite transformation*. in press (1970).

[13]  H. KANAMORI: *Velocity and Q of mantle waves*, in press, *Phys. Earth Planet. Int.* (1970).

[14]  H. KANAMORI and F. PRESS: *Nature*, **226**, 330 (1970).

[15]  V. I. KEILIS-BOROK and T. B. YANOVSKAYA: *Geophys. Journ.*, **13**, 223 (1967).

[16]  I. B. LAMBERT and P. J. WYLLIE: *Low velocity zone of the Earth's mantle: Incipient melting caused by water*, in press, *Science* (1970).

[17]  F. PRESS: *Journ. Geophys. Res.*, **73**, 5223 (1968).

[18]  F. PRESS: *Science*, **165**, 174 (1969).

[19]  F. PRESS: *Phys. Earth Planet. Int.*, **3**, 3 (1970).

[20]  F. PRESS: *Journ. Geophys. Res.*, **64**, 565 (1959).

[21]  F. PRESS: *Journ. Geophys. Res.*, **75**, 6575 (1970).

[22]  F. PRESS: *Symposium in honor of F. Birch*, E. ROBERTSON Editor, in press (New York, 1971).

[23]  D. FORSYTH and F. PRESS: *Journ, Geophys, Res.* (1971), in press.

[24]  C. WANG: *Journ. Geophys. Res.*, **75**, 3264 (1970).

[25]  L. R. JOHNSON: *Bull. Seismol. Soc. Amer.*, **59**, 973 (1969).

[26]  M. EWING, W. JARDETZKY and F. PRESS: *Elastic waves in layered media* (New York, 1957).

[27]  K. E. BULLEN: *Introduction to the Theory of Seismology* (Cambridge, 1963).

[28]  J. OLIVER: *Bull. Seismol. Soc. Amer.*, **52**, 81 (1962).

[29]  R. A. PHINNEY and D. L. ANDERSON: *Internal temperatures of the moon*, unpublished manuscript.

[30]  G. J. F. MACDONALD: *Journ. Geophys. Res.*, **67**, 2945 (1962).

[31]  F. BIRCH: *Physics of the crust*, in Sp. Paper 62, *Geol. Soc. Amer* (1955).

[32]  B. ISACKS, J. OLIVER and L. R. SYKES: *Journ. Geophys. Res.*, **73**, 5855 (1968).

[33]  F. J. VINE and H. H. HESS: *Sea Floor Spreading*, in *The Sea*, Vol. **4** (New York, 1970).

[34]  *Initial reports of the deep sea drilling project*, Vol. **3**, National Science Foundation, Washington, D. C. (1970.

[35]  M. TALWANI: *Marine Geol.*, **2**, 29 (1964).

[36]  L. R. JOHNSON: *Bull. Seismol. Soc. Amer.*, **59**, 973 (1969).

# The Inverse Problem of Seismology.

V. J. KEILIS-BOROK

*Institute of Physics of the Earth, Academy of Sciences of the USSR - Moscow*

## 1. – Introduction.

The purpose of these lectures is to outline the general framework of the modern methods of inversion of seismological observations into the Earth's structure; paying special attention to unsolved problems.

Section **2** contains the formulation of the problem; Section **3** describes an example of the solution of the idealized inverse problem: the inversion of travel-times of body waves, which are known completely as a piece-wise continuous function of time, and without the errors either in the travel-times, or in *a priori* idealization of the structure. This is the only variant of the inverse problem for which a complete and mathematically strict solution is obtained at the present time.

Even this idealized inverse problem has no unique solution: a set of different structures corresponds exactly to the same travel-time curve.

In the real situation the nonuniqueness will be larger, due to both kinds of errors mentioned above, and to the uncompleteness of the data. We have to respond to this challenge in two ways.

The first way is to reformulate the problem; to look for such properties of a structure, which can be determined from the given data; this way is outlined in Sect. **4**.

The second way is to invert jointly the whole set of the known relevant data; it should be done, of course, not instead of but together with the realistic formulation of the problem, according to Sect. **4**.

This way is described in Sect. **5, 6**. The basic framework of the method of inversion of real—uncomplete and unaccurate—data, and the specific problems, connected with this inversion, are outlined in Sect. **5**. The results of inversion of different sets of data by this method are described in Sect. **6**.

We do not go along the second way up to the end—to invert the seismic record as a whole, instead of separating it into the arrivals of different waves, and measuring their arrival times, amplitudes and velocities. The theory of such an inversion is not yet developed, though a one-dimensional problem is thoroughfully investigated [7].

## 2. – The formulation of the problem.

The inverse problem of seismology is to find the *internal* structure of the Earth from observations on its *surface*. More specifically, knowing the motion of a set of points on the Earth's surface, caused by some dynamic source we seek to find the distribution of elastic parameters $\lambda$, $\mu$, density $\varrho$ and dissipation constant $Q$ inside the Earth. The local velocities $\alpha$ and $\beta$ of longitudinal and transverse body waves, $\alpha = \sqrt{(\lambda + 2\mu)/\varrho}$, $\beta = \sqrt{\mu/\varrho}$, are usually considered instead of $\lambda$, $\mu$.

The set of points are seismological stations where recordings of the motion generated by an explosion or an earthquake source are taken over some frequency band.

The solution of this problem is nonunique because the number of stations is limited and because the information obtained from records is subject to errors. Moreover, it is not clear yet if there is a unique solution even in the abstract case of the absence of errors.

In order to reduce the nonuniqueness we have to impose some simplifications to the possible structure of the Earth. Thus we are looking for a *model* rather than for the real Earth, and our final formulation of the problem is as follows: Knowing some properties of seismic waves measured in a set of points on the Earth's surface and assuming some hypotheses about the Earth's structure we seek to find a set of Earth models, fitting these hypotheses and the given data, making allowance for the precision of the data.

This problem is as old as seismology itself, and many excellent results —such as the discovery of the Earth's crust and the Earth's core—were obtained by comparatively simple methods, whereas the formulation of the problem remained essentially intuitive. Why, therefore, do I try to formalize it? I hope the answer will be found in the lectures. In short the answer is: because we are using computers and looking for the fine details of the Earth's structure.

## 3. – Inversion of a complete and exact travel-time curve.

Let us investigate a theoretical sub-problem: what information can be obtained from a travel-time curve in the ideal case of a spherically symmetric Earth and an exactly known curve. This problem was solved in 1907 (for references see [9]), under the assumption that no wave guides are present in the Earth. The solution in the presence of wave guides was obtained in 1966 [5] and I will follow the treatment of that work here. The preliminary information on the theory of travel-time curves can be found in [9].

3'1. *The problem*. – The travel-time curve $t(\varDelta)$ for a body wave, propagating in a sphere $0 \leqslant r \leqslant R$ from a point source on the surface $r = R$ is given, $\varDelta$ being the epicentral distance in radians. And the following assumptions are made:

i) The wave is propagated according to the laws of geometrical optics.

ii) The sphere is spherically symmetric, *i.e.* the velocity $v$ ($v = \alpha$ or $\beta$) depends on $r$ only: $v = v(r)$.

iii) The nature of the rays, for which travel-times are given, is the following: They return to the surface for the first time (we do not consider surface reflections). If the ray meets a discontinuity, it splits into two reflected and two refracted rays; we consider only rays of the waves of the same nature as the incident wave—longitudinal or transverse; we take the refracted ray provided it exists and otherwise, (in the case of total reflection) the reflected ray; for a branching ray we take the upper branch.

Three additional assumptions are introduced for formal reasons:

iv) $v(r)$ and its first and second derivatives are continuous for all $r$, except for a finite number of points.

v) $v(r)$ is positive and limited.

vi) The number of wave-guides is finite (the strict definition of a wave-guide is given below).

Evidently, these assumptions do not impose on $v(r)$ any noticeable limitations.

The problem is *to find $v(r)$ from $t(\varDelta)$*. We cannot know $\varDelta$ beforehand; we know the distance $\tilde{\varDelta}$ between the source and a point, where the ray returns to the surface; the ray, however, can turn around the centre of the Earth before reaching the surface.

Strictly speaking we, therefore, only know

$$\tilde{\varDelta} = \varDelta - 2k\pi \, , \qquad\qquad k = 0, 1, 2, \dots \, ,$$

with $k$ unknown.

It is usually assumed that $k = 0$, since for larger $k$ the intensity of the wave would be too small. But in studying the core, for example, we have to be careful.

Hence, our problem has two parts: firstly to find $t(\varDelta)$ from $t(\tilde{\varDelta})$ and then secondly to find $v(r)$ from $t(\varDelta)$. We consider here the second part, evidently the main one; the treatment of the first part can be found in [5, 6].

**3'2.** *The form of the equation for, and some properties of, the travel-time curve.* –

*Equation.* The assumptions i), ii) lead to the following formulae for the travel-time curve [9]:

$$(1) \qquad t(\varDelta) = \int_s \frac{ds}{v} ,$$

$$ds = \sqrt{dr^2 + r^2 \, d\varDelta^2} ,$$

$$(2) \qquad \sin i(r) = p \frac{v(r)}{r} R .$$

Here $s$ is the seismic ray; its equation, under assumption ii) is (2). $i(r)$ is the angle between the ray and the radius $r$, $p$ is the « ray parameter »—it is constant along the ray.

One can think of $p$ as representing either the inclination of the ray at the source $p = \big(1/v(R)\big) \sin(s, R)$ or the velocity at the lowest point of the ray, $r = r_{\min}$, where $i = \pi/2$,

$$p = \frac{r_{\min}}{R v(r_{\min})} .$$

$p$ can be determined from the known travel-time curve: $p = dt/d\varDelta$. It is more convenient to reduce our problem to the one on a half-plane.

The transformation,

$$(3) \qquad x = \frac{R\varDelta}{v(R)} ; \qquad y = \frac{R}{v(R)} \ln \frac{R}{r} ; \qquad u = \frac{v(r)}{v(R)} \frac{R}{r} ,$$

transforms our sphere into the half-plane $y \geqslant 0$, with a travel-time curve $t(x) = t[\varDelta(x)]$ and ray equation

$$(4) \qquad \sin i(y) = p/u(y) .$$

Our problem is now to find $u(y)$ from $t(x)$; then we can retransform $u(y)$ into $v(r)$.

Let us determine the analytical relation between $t(x)$ and $u(y)$. The simplest way is to notice that $x$ and $t$ have the following increments when $y$ increases to $y + dy$:

$$dx = \operatorname{tg} i(y) \, dy , \qquad dt = dy/u(y) \cos i(y) .$$

Substituting $i$ from (4) in these increments and integrating we obtain the well-known parametric equation for $t(x)$,

$$(5) \qquad x(p) = 2 \int_0^{Y(p)} \frac{p \, dy}{\sqrt{u^{-2} - p^2}} , \qquad t(p) = 2 \int_0^{Y(p)} \frac{dy}{u^2 \sqrt{u^{-2} - p^2}} .$$

$Y(p)$ is the lowest depth reached by the ray with parameter $p$; in other words $Y(p)$ is the smallest depth at which (4) gives imaginary $i$,

(6)                           $Y(p) = \inf \{y, \; pu(y) \geqslant 1\} \,,$                           $p \in (0, 1) \,.$

Here and later « inf » (infimum) and « sup » (supremum) mean the lower and the upper limits. $Y(p)$, as is evident from (6), does not increase with $p$: the steeper leaves the ray the source, the deeper it penetrates. The parameter $p$ decreases monotonously along the travel-time curve (starting from the source $t = 0$, $x = 0$, where $p = 1$).

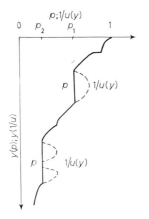

$Y(p)$ is related to $u(y)$ in the way shown in Fig. 1: $Y(p)$ is a function which is the inverse of $p(y) = 1/u(y)$ away from the minima of $u(y)$; and in any interval where $u(y)$ takes a value less than at any shallower depth, $Y(p)$ becomes a vertical segment. Such intervals of depth represent the wave-guides. It is their possible presence which makes the inversion of travel-times nonunique and interesting.

*Wave-guides.* The formal definition of wave-guide is the following:

Denote

$$s(y) = \sup \{u(y^0), \qquad y^0 \in [0, y)]\} \,.$$

Fig. 1. – Definition of the function $Y(p)$ (solid line), after [5]. Dashed lines: the structure $1/u(y)$ at the wave-guides; outside the wave-guides $1/u(y)$ coincides with $Y(p)$.

The wave-guides are such intervals of the depth $y$,

$$y_k < y < \bar{y}_k \,, \qquad k = 1, 2, \dots, \bar{k}, \qquad \bar{y}_k < y_{k+1}$$

for which the following conditions hold:

  inside each interval $s(y) = \mathrm{const}$;

  each interval contains the points, where $u(y) < s(y)$;

  outside these intervals $u(y) = s(y)$.

Evidently, for continuous $u(y)$, $u(y_k) = u(\bar{y}_k)$.

It is easy to see from (4), that no ray has a minimum inside the wave-guides. The ray with parameter $p = p_k + 0$, $p_k = 1/u(y_k)$, has a deepest point $y_{\min} = y_k - 0$, just above the wave-guide; the « next » ray, with $p = p_k - 0$, has a deepest point $y_{\min} = \bar{y}_k + 0$, just below the wave-guide. That is why in the presence of wave-guides the travel-time curve shows a special discontinuous pattern (Fig. 2): if we follow the travel-time curve from $(t = 0,\ x = 0)$ onwards it will terminate at the point $t(p_k + 0)$, $x(p_k + 0)$, and then appear again at the point $t(p_k - 0)$, $x(p_k - 0)$. Naturally, $t(p_k - 0) - t(p_k + 0) > 0$, $x(p_k - 0) - x(p_k + 0) > 0$ (these differences are the time and distance, which are spent in passing the wave-guide).

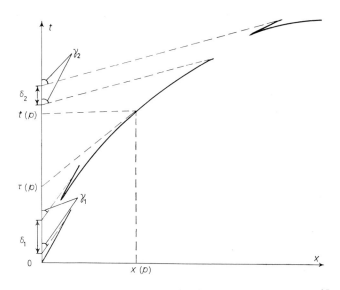

Fig. 2. – Discontinuity of travel-time curve in the presence of wave-guides. $\operatorname{ctg} y_k = p_k$.

One integral property of the wave-guide is characterized directly by the jump of the tangent to $t(x)$ at this discontinuity. Let us draw a tangent to $t(x)$ at some point $t = t(p)$. Denote $\tau(p)$ the co-ordinate on the $t$-axis, at the intersection with this tangent (Fig. 2),

$$(7) \qquad\qquad \tau(p) = t(p) - pX(p) .$$

It follows directly from (5), that

$$(8) \qquad \sigma_k = \tau(p_k - 0) - \tau(p_k + 0) = \int_{y_k}^{\overline{y}_k} \sqrt{u^{-2} - p_k^2}\, dy .$$

This means that the number of wave-guides $\overline{k}$ and their integral properties $\sigma_k$ are known exactly from exact $p(x)$, $\sigma_k$ or $t(x)$ or $\tau(p)$ curves.

The following properties of the function $\tau(p)$ will be used later:

a) $\tau(p)$ decreases monotonously with increase of $p$.

b) $\tau(p)$ is continuous, except the points $p = p_k$, where it has an instantaneous increment $-\sigma_k$, given by (8)

(9)    (c)
$$\frac{\partial \tau}{\partial p} = -X(p) - \sum_k \sigma_k \delta(p - p_k),$$

where $\delta$ is Dirac's delta-function. The sum is taken over all $k$, for which $p_k > p$.

3'3. *The solution.* – Our problem can now be formulated analytically: to find $u(y)$ from eqs. (5), where $t(p)$, $x(p)$ are known.

It is found in [5], that the velocity $u(y)$ *outside* the wave-guides can be determined uniquely from $t(x)$, *if* the wave-guide intervals $(y_k, \bar{y}_k)$ and $u(y)$ inside them are known.

However, the wave-guides are determined by $t(x)$ nonuniquely; consequently, there is a set of different $u(y)$, which correspond exactly to the same $t(x)$.

The exact solution of our inverse problem is formulated in the following theorem, which is proved in [5].

Suppose we found such wave-guide intervals $y_k \leqslant y \leqslant \bar{y}_k$ $k = 1, 2, ..., \bar{k}$ and a velocity-depth distribution $u(y)$ within them, so that the following three conditions are satisfied:

a) The function

(10)
$$Y(q) = \Phi(q) + \Psi[q, u(y)],$$

does not increase at $0 \leqslant q \leqslant 1$.

Here

(11)
$$\Phi(q) = \frac{2}{\pi} \int_q^1 \frac{X(p)\,dp}{\sqrt{p^2 - q^2}}$$

and

(12)
$$\Psi(q) = \frac{2}{\pi} \sum_{i=1}^k \int_{y_i}^{\bar{y}_i} \mathrm{arc\ tg} \left[ \frac{\sqrt{u^{-2} - p_i^2}}{\sqrt{p_i^2 - q^2}} \right] dy,$$

$$\text{for } p_{k+1} < q < p_k, \ 1 \leqslant k \leqslant \bar{k}.$$

$$\Psi = 0, \qquad\qquad\qquad \text{for } q > p_1.$$

b)
$$Y(p_k + 0) = y_k, \qquad Y(p_k - 0) = \bar{y}_k,$$

c)
$$\int_{y_k}^{\bar{y}_k} \sqrt{u^{-2} - p_k^2}\,dy = \sigma_k.$$

Let us determine $u(y)$ outside the wave-guides from the function $Y(q)$, given by (10) with $X(p)$, $\delta_k$, $\bar{k}$ determined by the given $t(x)$; we assume, as in Subsect $3\cdot 2$, that $Y(p)$ is the depth, at which $u = 1/q$; a strict definition is given again by (6). We have so determined $u(y)$ on the whole range of depths. The theorem is that all such and only such $u(y)$ fit the given $t(x)$ exactly; in the absence of wave-guides $\Psi = 0$ and (10) coincides with Herglotz-Wiechert's formula.

For a given $\delta_k, p_k$ a whole set of the wave-guides boundaries $(y_k, \bar{y}_k)$, $k = 1, 2, ..., \bar{k}$, and the velocities in them, $u(y)$ for $y_k < y < \bar{y}_k$, fit the above conditions.

From (10), (12), we see that different wave-guides can give different velocities outside the wave-guide. So, the set of structures corresponding exactly to the same travel-time curve in the presence of wave-guides is the solution of our problem (just as the set $x = 1$, $x = -1$ is the solution of the problem $x^2 = 1$).

We now have to compute this set and describe its general properties.

We describe first the area covered by all possible structures in the $(u, y)$ plane (Fig. 3); this is such an area that all structures are inside it and at least one structure passes through each of its points. However, not every curve which may be drawn in this area is a possible structure. An upper limit for $u$ is $\Phi(p)$, determined by the Herglotz-Wiechert formula; it may be used in the presence of wave-guides, since $X(p)$ is determined on all the range of integration. So, $\Phi(p)$ has the physical interpretation of being the upper bound for $u$ at a depth $Y(p)$. The lower bound has the form shown in Fig. 3 for two wave-guides; the authors of [5] assumed that number to make the Figure giraffe-like.

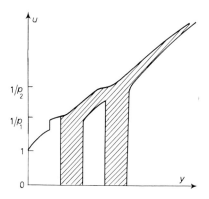

Fig. 3. – The nonuniqueness in the presence of two wave-guides for surface source, after [5].

Up to the first wave-guide we have a unique structure, and we also know the depth $y_1$ of the top of the first wave-guide. From this point on the problem becomes computational: we construct a set of structures for a given shape of wave-guide.

For example, let us consider a rectangular shape: a constant velocity inside the wave-guide. Then for each value of the velocity we compute the thickness of the wave-guide by formula (8) and test them against condition a). If this condition is satisfied we have obtained one of the possible wave-guides, and then, by (10), one possible structure $u(y)$. It can be proved that a sufficiently narrow wave-guide always satisfies $a$): we may take a very small velocity (*close to zero*) and compute the corresponding thickness which will be vanishingly small; and the $u(y)$ below such wave-guide will lay close to the upper boundary, corresponding to $\Phi(p)$.

It is, therefore, impossible to impose a lower limit on the velocity in a wave-guide by pure travel-time data; condition $a$) will be violated, however, when the velocity drop in the wave-guide becomes too small; this gives the maximum thickness and velocity of the wave-guide if its shape is rectangular.

The same procedure can be applied to any other shape. We can invert a given $t(x)$ by this technique into a set of structures $u(y)$ with a given shape of wave-guides. By the same technique a given structure can be inverted into a set of equivalent structures which give exactly the same $t(x)$. In particular, as was first pointed out by SCHLICHTER, equivalent wave-guides may be obtained by reshuffling their layers. More generally, structures are equivalent if they are represented by the same functions

$$(13) \qquad\qquad F_k(\overline{u}) = \text{mes }\{y, \ y < y_k, \ u(y) < \overline{u}\};$$

$F_k(\overline{u})$ is the total length of the intervals in which the velocity is less than $\overline{u}$. As an exercise one can compute in this way the uncertainty in the radius of the Earth's core, generated by wave-guides in the mantle.

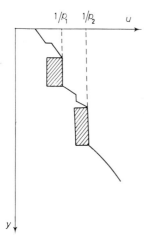

Fig. 4. – The nonuniqueness in the case, when $t(x)$ is known for the surface source and for sources between the wave-guides. After [5, 16].

**3'4.** – *The travel-time curve $t(x)$ from a deep focus* gives additional information [5, 6, 16]:

$a$) The focal depth $h$—independently of the structure; this possibility seems unexpected.

$b$) $\ F_h(\overline{u}) = \text{mes }\{y, \quad y < h, \quad u(y) < \overline{u}\},$

the total length of the interval above the source, where the velocity is less than any given $\overline{u}$.

$c$) If the source is between the wave-guides we can determine *uniquely* the depth of their boundaries nearest to the source and the velocity distribution $u(y)$ between them. With information from the sources above and below a particular wave-guide we can also determine a lower limit for the velocity in this wave-guide.

So, if we know the travel-time curve from sources located above the upper wave-guides and also between each wave-guide, then the area of nonuniqueness, shown on the Fig. 3, is reduced to the area shown schematically in Fig. 4.

*The travel-time curve for a reflected wave* has similar properties; in this case $h$ would be the depth of a reflecting boundary [16].

The computational methods to use all these possibilities have not yet been developed.

3'5. *The travel-time curve consisting of a finite number of points.* – The rays, radiating from the source in some range of the angles $i$, may return to the surface at the same distance. It means, that for some structures $u(y)$ the function $X(p)$ may be constant for some range of $p = \sin i(R)$.

It is easy to find some examples of such structures from (10), (11), in the absence of wave-guides. Suppose, for example, that the travel-time curve consists of two points only: all rays with $p \leqslant p_1$ emerge at $x = c$, and all rays with $p > p_1$ emerge at $x = d$.

Substituting this into (8), (9) we will find a structure which is surprisingly simple for such a fancy travel-time curve,

$$u(y) = \cosh\left(\frac{2c}{\pi} y\right), \qquad\qquad \text{for } 0 \leqslant y \leqslant y_1;$$

$$u(y) = \frac{1}{p_1} \cosh\left[\frac{2d}{\pi}(y - y_1)\right], \qquad\qquad \text{for } y > y_1;$$

$$y_1 = \frac{\pi}{2c} \operatorname{arccosh} \frac{1}{p_1}.$$

More general statement [6]: A set of points $(t_i, x_i)$, $i = 1, 2, ..., Z$, $x_1 < x_2 < ... < x_Z$ is a degenerated travel-time curve for some structure $u(y)$, if and only if $x_1 \leqslant t_1 < t_2 < , ..., < t_Z$.

This result is fancy, but not miraculous, since the gap between isolated focusing points $(t_i, x_i)$ will be filled in by diffracted waves, with straight line travel-time curves.

## 4. – Inversion of uncomplete exact data.

4'1. – *The possible limits for the structure* (from the $t$, $dt/dx$ at a final set of distances $x$).

All methods of inversion of the travel-times assume the knowledge of a *continuous* travel-time curve $t(x)$.

Practically, this curve is interpolated between observed points; interpolation is always nonunique; this nonuniqueness becomes essential if the real travel-time curve has loops and/or discontinuities, connected with the wave-guides. That is why the following problem arises: what information can be obtained from the observed data, as they are, without interpolation?

This problem, in spite of its fundamental nature for seismology, was solved (with some simplifications) only recently [3]; I will describe schematically the basic results of [3].

*The formulation of the problem.* A finite set of the values $(p_i, t_i, X_i)$, $i = 1, 2, ... n,$, is given, $t_i$ being the travel-time and $p_i = dt/dX$ at the distance $X_i$; $p_i < p_{i+1}$, $p_{n+1} = 1$. The same assumptions, as in Subsect. 2·1 are introduced, including the absence of errors; and we consider the problem, already reduced to the plane, by transformation (3), so that formulae (4)-(10) hold. We introduce also two additional assumptions which are physically trivial:

vii) The velocity $u(y)$ in the wave-guides is not less than some constant.

viii) $X(p)$ has a finite number of infinite branches, *i.e.* $du/dy$ has a finite number of zero's.

It is more convenient to investigate not the velocity-depth function $u(y)$ but the inverse function $y(s)$, where $s = 1/u$ — « the slowness ».

Our problem is to find the upper and lower limits for the average value of $y$ on some interval of $s$, *i.e.* to find such $\bar{Z}_j$, $\underline{Z}_j$, that

(14)                        $$\bar{Z}_j \geqslant Z(a_j, b_j) \geqslant \underline{Z}_j ,$$

where

(15)                 $$Z(a_j, b_j) = \frac{1}{b_j - a_j} \int_{a_j}^{b_j} y(s) \, ds , \qquad j = 1, 2, ..., \bar{j} \leqslant n.$$

The replacement of the traditional presentation of the structure, $u(y)$, by $y(s)$ needs no justification: $y(s)$ and $u(y)$ are in principle equivalent. However, $y(s)$ is more adequate to the nature of the problem: if at one single point $dt/dX = s$ is directly means (see (4)) that at some depth $y$ the slowness is equal to $s$; and it is the determination of this depth $y$ which is nonunique and demands the knowledge of the whole travel-time curve.

The replacement of $y(s)$ by the averaged structure $Z(a, b)$ needs a justification, since we apparently step back from the « ideal goal »—the determination of the « structure ». The justification is the following (see also [2]): the results, described in Sect. **3**, show that this determination is unstable: little variation of $t(X)$ may lead to the big change of $u(y)$. This can be illustrated by an example, suggested by PJATEZNI-SHAPIRO and shown on Fig. 5.

Solid lines represent a travel-time curve $t(X)$ with discontinuities, created by the wave-guides, and a corresponding structure $u(y)$ with deep narrow wave-guides.

Dashed lines represent a continuous travel-time curve $t_0(x)$. It corresponds to some monotonous structure $u_0(y)$, which may obtained by HERGLOTZ-WIECHERT formulae ((10)-(12) with $\Psi = 0$). What happens when the amplitudes $\delta_k$ of discontinuities of $t(X)$ decrease so that $t(X) - t_0(X) \to 0$? The

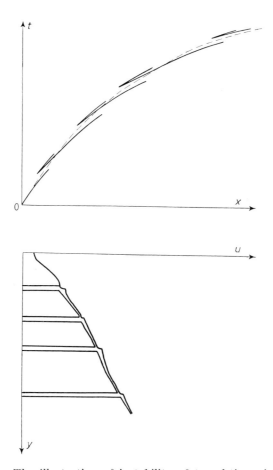

Fig. 5. – The illustration of instability of travel-times inversion.

difference of structures $u(y) - u_0(y)$ will *not* decrease with $\sigma_k$; according to Sect. 2 (see Fig. 3) $u(y)$ can be close to zero in each wave-guide, independently of $\sigma_k$.

The thickeness of the wave-guides, however, will decrease with $r$, according to (8); consequently, only the determination of smoothed, somehow averaged,

structure may be stable. The advantage of investigation of the averaged structure $Z(a_j, b_j)$ instead of $u(y)$ or $y(s)$ is now clear. Of course, when the difference
between $p_i$ in the points of observation decreases, the intervals of averaging
can be decreased too.

*The solution.* The averaged structure $Z(a_j, b_j)$ outside the wave-guides can
be represented in the following form (by substituting (10) into (16)):

$$(17) \qquad\qquad Z(a, b) = I(a, b) + J(a, b) ,$$

where $I$ and $J$ are the averaged functions $\varPhi$ and $\varPsi$,

$$(18) \qquad\qquad I(a, b) = \frac{1}{b-a} \int_a^b \varPhi(q)\,\mathrm{d}q ,$$

$$(19) \qquad\qquad J(a, b) = \frac{1}{b-a} \int_a^b \varPsi(q)\,\mathrm{d}q .$$

Substituting (11), (7), (9) into (18) and integrating, we have [3]

$$(20) \qquad\qquad I(a, b) = I_1(a, b) - I_2(a, b) ,$$

where

$$I_1(a, b) = \frac{2}{\pi(b-a)} \int_0^1 \tau(p)\, \frac{\partial\varphi(p, a, b)}{\partial p} ,$$

$$I_2(a, b) = \frac{2}{\pi(b-a)} \sum_k \sigma_k \varphi(p_k, a, b) .$$

The sum is taken over all $k$, for which $a \leqslant p_k \leqslant 1$.

$$(21) \qquad \left| \begin{array}{ll} \varphi = \arccos \dfrac{a}{p}, & \text{for } a \leqslant p < b, \\[2ex] \varphi = \arccos \dfrac{a}{p} - \arccos \dfrac{b}{p}, & \text{for } b \leqslant p \leqslant 1. \end{array} \right.$$

The limits for $I(a, b)$ and some information on possible wave-guides can be
obtained from (20) and the properties of the function $\tau(p)$, indicated in Subsect. 3˙2.

Evidently,

$$(22) \qquad \frac{2}{\pi(b_j - a_j)} \int_a^1 \bar{\lambda}_j(p) \frac{\partial \varphi}{\partial p} \, dp > I_1(a_j, b_j) > \frac{2}{\pi(b_j - a_j)} \int_a^1 \underline{\lambda}_j(p) \frac{\partial \varphi}{\partial p} \, dp \,,$$

if

$$(23a) \qquad \bar{\lambda}_j(p) \geqslant \tau(p) \,, \qquad \underline{\lambda}_j(p) \leqslant \tau(p) \quad \text{on } a_j \leqslant p < b_j, \text{ where } \frac{\partial \varphi}{\partial p} > 0$$

and

$$(23b) \qquad \bar{\lambda}_j(p) \leqslant \tau(p) \,, \qquad \underline{\lambda}_j(p) \geqslant \tau(p) \quad \text{on } b_j \leqslant p \leqslant 1 \,, \text{ where } \frac{\partial \varphi}{\partial p} < 0 \,.$$

Now we have to estimate $\bar{\lambda}_j(p)$, $\underline{\lambda}_j(p)$ from the given data $(p_i, t_i, X_i)$. We assume $a_j = p_j$, $b_j = p_{j+1}$, and investigate the average structure $Z_j = Z(p_j, p_{j+1})$ (14) on each interval

$$(24) \qquad p_j \leqslant p \leqslant p_{j+1} \,, \qquad j = 1, 2, \ldots, n \,.$$

Let us represent the given data on the $(p, \tau)$-plane, with $\tau(p_j) \equiv \tau_j = t_j - p_j X_j$ (Fig. 6). Since $\tau(p)$ is a monotonous function, it lays inside the chain of squares on Fig. 6. Consequently, $\underline{\lambda}_j(p)$, $\bar{\lambda}_j(p)$ can be drawn along the boundaries of this chain. According to (23),

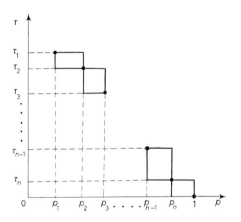

Fig. 6. – The limits for $\tau(p)$, after [3].

$$(25) \qquad \begin{vmatrix} \bar{\lambda}_j(p) = \tau_j \,, & \underline{\lambda}_j(p) = \tau_{j+1} & \text{for } p_j \leqslant p \leqslant p_{j+1} \,, \\ \bar{\lambda}_i(p) = \tau_{i+1} \,, & \underline{\lambda}_i(p) = \tau_i & \text{for } p_i \leqslant p \leqslant p_{i+1} \,, \qquad j < i \leqslant n \,. \end{vmatrix}$$

We have also the following limitation for the wave-guides in the interval $p_j < p < p_{j+1}$,

$$\sum_k \sigma_k = \tau_{j+1} - \tau_j \,.$$

The sum is taken over all wave-guides within this interval. The limits for $I_1$ can be made much more narrow, if we introduce an additional assumption:

ix) At some (known) distances the travel-times are given for all branches of the travel-time curve.

This hypothesis, together with (9), gives the limits for $d\tau/dp$.

Let us find for each interval $p_j < p < p_{j+1}$ two distances, $\underline{X}_j$ and $\overline{X}_j$, from the following conditions: all arrivals at these distances have $p$ outside the interval $(p_j, p_{j+1})$,

$\underline{X}_j$ is the largest of these distances, which is less than $X_j$ and $X_{j+1}$;

$\overline{X}_j$ is the smallest of these distances, which is larger than $X_j$ and $X_{j+1}$.

Then, according to (9),

$$(26) \qquad\qquad -\overline{X}_j < \frac{d\tau}{dp} < -\underline{X}_j, \qquad\qquad \text{for } p_j \leqslant p \leqslant p_{j+1}$$

except the points of discontinuity of $\tau(p)$.

The further scheme of definition of the limits $\underline{\lambda}_j$, $\overline{\lambda}_j$ is shown on Fig. 7. We draw through the point $(p_j, \tau_j)$ two straight lines to the right; tg $\bar\eta = \overline{X}_q$, tg $\eta = \underline{X}_q$. These lines will intersect the line $p = p_{k+1}$ at some points, with ordinates $\nu$ and $\mu$.

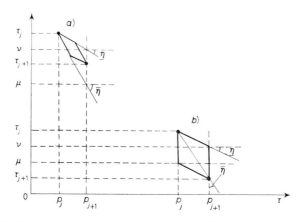

Fig. 7. – The limits for $\tau(p)$ under assumption (ix), after [3]. a) $\tau_{j+1} \geqslant \mu$, wave-guides are possible; b) $\tau_{j+1} < \mu$, wave-guides exist.

There are two possibilities:

$$\nu > \tau_{j+1} \geqslant \mu \quad \text{(Fig. 7 }a\text{))} \qquad \text{and} \qquad \mu > \tau_{j+1} \quad \text{(Fig. 7 }b\text{))}.$$

If $\tau_{j+1} \geqslant \mu$ we draw the lines with the same slopes through the point $(\tau_{j+1}, p_{j+1})$ to the left. $\tau(p)$ lays inside the parallelogram formed by these two pairs of lines (Fig. 7 a)).

The wave-guides in this interval fit the limitation

$$(27) \qquad\qquad 0 \leqslant \sum \sigma_k \leqslant \nu - \tau_{j+1}.$$

Possibly the wave-guides do not exist in this range; if they do exist, the sum of their integral characteristics (8) does not exceed $v - \tau_{j+1} > 0$.

If $\tau_{j+1} < \mu$ we draw through the point $(p_{j+1}, \tau_{j+1})$ the line with the smallest slope, $X_j$; $\tau(p)$ lays inside the parallelogram, shown on Fig. 7 $b$).

The wave-guides in this interfal $do$ exist and fit the limitation

$$(28) \qquad \tau_{j+1} - \mu \leqslant \sum \sigma_k \leqslant v - \tau_{j+1} .$$

The parallelograms, shown on Fig. 7 $a$), $b$) form a continuous chain between $\tau_1$ and $\tau_{n+1} = 0$. $\tau(p)$ lays inside this chain and we have an estimation of $\underline{\lambda}_j$, $\bar{\lambda}_j$: more exact than by (25).

For $p_j < p < p_{j+1}$, $\bar{\lambda}_j$ is the upper, $\underline{\lambda}_j$ is the lower boundary of the parallelogram between the points $(\tau_j, p_j)$ and $(\tau_{j+1}, p_{j+1})$. For the rest of the $p$-th interval, $1 \geqslant p \geqslant p_{j+1}$, $\bar{\lambda}_j$ is the lower, and $\underline{\lambda}_j$ the upper boundary of the chain of parallelograms. The limits for $I_1$ can be obtained now from (22). The formulae (27), (28) give also the limits for $I_2$. The final explicit expression for the limits of $I = I_1 + I_2$ can be found in [3].

It is possible now to estimate the second term in (17), and therefore, the limits for the averaged structure $Z_j$ outside the wave-guides; simultaneously we can estimate the thickness of the wave-guides and the velocity in them.

This possibility is clear from the following simplified example: let us replace the wave-guides in each of the intervals (24) by one equivalent wave-guide, with a thickness $h_j$ and a slowness $s_j$. According to (10)-(12), (17)-(19), (27), (28), $h_j$, $s_j$ should satisfy the following conditions

$$(29) \qquad \begin{cases} a) \quad \max\left(0, \dfrac{\tau_{j+1} - \mu}{\sqrt{s_j^2 - p_j^2}}\right) \leqslant h_j \leqslant \dfrac{v - \tau_{j+1}}{\sqrt{s_j^2 - p_{j+1}^2}}, \qquad j = 1, 2, ..., n-1; \\[3mm] b) \quad I(p_j, p_{j+1}) + J(p_j, p_{j+1}) \end{cases}$$

do not increase with the decrease of $j$. The second condition is the consequence of the fact, that $Y(q)$ does not increase with $q$, what simply means that the velocity is a single-valued function of depth.

Since we assumed that the upper limit for $s_j$ is known (see vii) p. 252), the condition $a$) determines the lower limits for $h_j$. Condition $b$) determines the upper limits for $h_j$ and, consequently, the lower limits for $s_j$ (see (8)).

The computational scheme for estimation of the limits for $s_j$, $h_j$, as well as the limits for $Z_j$ when the velocity in the wave-guides is not necessarily constant, is given in [3]. Figure 8 illustrates what kind of results can be obtained by this method. It shows the numerical experiment with the $S$-wave velocity-depth distribution found in [8] for the Western United States.

A set of the values $(p_i, t_i, X_i)$, computed in [8] for this distribution was

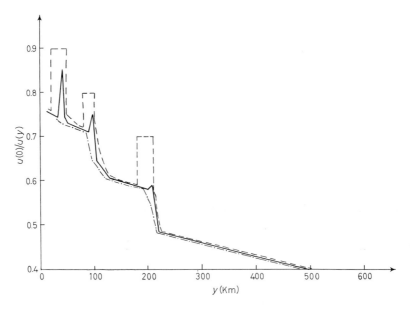

Fig. 8. – The limits for the structure from $\tau(p)$, corresponding to the shear-wave velocity model, given in [8]. Dashed lines: the limits, solid line: the « true » model. After [3].

taken, as « initial data », and the corresponding velocity limits were calculated by the method described. These limits and the initial structure are shown on Fig. 8. For details, see [3].

**4˙2. *The resolving power of the data.*** – The contradiction between discrete nature of observations and continuous nature of the structure we are looking for, is more general than it was formulated at the beginning of Subsect. 4˙1.

Sometimes this smoothing is introduced indirectly, by drawing the smoothed curves through the cloud of the measured values; sometimes directly as in inverting the free oscillations. This smoothing of the structure determines the exact meaning of the seismological models of the Earth; amazingly however, it remained alsmost completely univestigated, except the few attempts related to $t(\varDelta)$, $A(\varDelta)$ where $A$ means the amplitude [1, 10].

That is why I think that one of the major break-throughs in the seismology in recent years was made in [2], by formulating the following problem: which properties of an Earth model $\alpha(r)$, $\beta(r)$, $\varrho(r)$, $Q(r)$ can be uniquely determined by a given set of observed data? This problem was solved in [2] for a linear approximation and I shall describe its basic idea on a simple example. Suppose that some data, for example, — some periods of free oscillation $T_k$, are known exactly and that we also know an initial Earth model, say $\alpha_0(r)$, $\beta_0(r)$, $\varrho_0(r)$, $Q_0(r)$ for which the calculated free periods $T_{k_0}$ are very close to the known ones:

suppose also that $T_{k_0}$ is practically constant for possible perturbations of all parameters, except, say, $\beta$. In this case we have the linear relation

$$(30) \qquad T_k - T_{k_0} = \int_0^R [\beta(r) - \beta_0(t)] \, \mathscr{G}_k(r) \, dr \,.$$

Taking a linear combination of the $(T_k - T_{k_0})$ we have

$$(31) \qquad \sum_k a_k(\bar{r})(T_k - T_{k_0}) = \int_0^R [\beta(r) - \beta_0(r)] A(r, \bar{r}) \, dr \,,$$

where

$$(32) \qquad A(r, \bar{r}) = \sum_k a_k(\bar{r}) \, \mathscr{G}_k(r) \,.$$

The resolving power of our data $T_k$ for $r = \bar{r}$, i.e. for a depth $(R - \bar{r})$ can be measured by the difference between $A(r, \bar{r})$ and the delta function $\delta(r - \bar{r})$ since $A(r, \bar{r}) = \delta(r - \bar{r})$ means perfect resolution.

An integral

$$\int_0^R A^2(r, \bar{r})(r - \bar{r})^2 \, dr$$

is used in [2] as a measure (among others) of this difference. The coefficients $a_k$ are to be chosen in such a way as to minimize this measure, maintaining the natural normalization

$$(33) \qquad \int_0^R A(r, \bar{r}) \, dr = 1 \,.$$

The function $A(r, \bar{r})$ thus obtained gives an averaging kernel, which is as narrow as possible for the given data. Only such details in $\beta(r) - \beta_0(r)$ which do not disappear after smoothing with this kernel $A$ may be determined.

This idea is generalized in [2] for the case when the model is determined by several parameters and also the observations contain errors with a known correlation matrix. This method would solve the inverse problem in some sense uniquely, under condition that the linear approximation analogous to (30) were true for our data in the whole possible range of $\alpha, \beta, \varrho, Q$. Unfortunately this condition is satisfied only for surface waves and free oscillations on sufficiently long periods. If this condition does not hold, the described approach will give for different models of the Earth some different sets of corrections, smoothed in the same way.

The application of this method to the travel-time data was, up to now, unsuccessful: for the structure in the form (5) the problem is essentially non-linear, due to the presence of $Y(p)$ in the limit of integration.

The transformation, suggested in [3], $dy = y'(s)\,ds$ makes the problem linear *even without the knowledge of the initial model*; for example, in the absence of wave-guides we have

$$(34) \qquad\qquad \tau(p) = 2 \int_{p}^{1} y(s)\, \frac{s\,ds}{\sqrt{s^2 - p^2}}.$$

This kernel, however, does not allow to get a good resolving power.

## 5. – General framework for the inversion of real observations.

5‘1. *Two schemes of inversion* (*after* [11, 13]). – Two general schemes of inversion are given in Fig. 9. The upper scheme describes the search for solu-

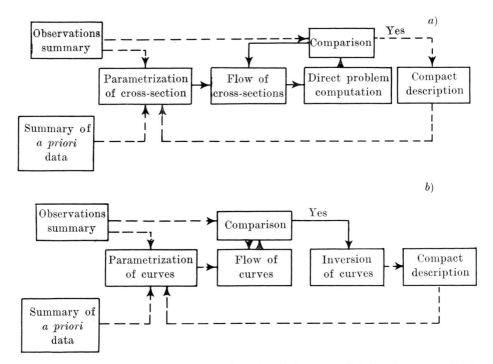

Fig. 9. – The general scheme of inversion of real data. *a*) Trial and error method, *b*) direct inversion metod. Dashed lines: once-round operations, solid lines: repetitive cycles. After [11].

tions in the space of unknown structures. The structure is determined as a set of functions $\alpha(r)$, $\beta(r)$, $\varrho(r)$, $Q(r)$; all or some of them can be investigated simultaneously, depending on the data. The general scheme will not change, if $\alpha$, $\beta$, $\varrho$, $Q$ depend on all three spacial co-ordinates. But the corresponding know-how is still not developed, and the sense of humour should prevent us from replacing $r$ by $(r, \varphi, \theta)$ in this context.

The lower scheme describes the search in the space of the regularities in the seismic wave propagation. These are the same regularities which we determine from the observations: the travel-time curve $t(\varDelta, h)$; its derivative $dt/d\varDelta$; the amplitude-distance curve $A(\varDelta, h, T)$; its derivatives $\partial A/\partial \varDelta$, $\partial A/\partial T$; the ratio of the amplitudes for different components (*i.e.* the polarization); the phase-and group-velocity of surface waves $c(T)$, $U(T)$; the periods of free oscillation $T_k$ and their attenuation $\lambda_k$. All or some of these functions and numbers can be used simultaneously.

The first method was called the trial-and-error method, the second the method of direct inversion.

*The trial-and-error method* is basically the following. At first the unknown cross-section is replaced by a set of parameters. More specifically, it is presented as a function of $r$ depending on these parameters. The type of function is fixed and the determination of the cross-section is reduced to the determination of numerical values of the parameters. The possible limits of these parameters, *i.e.* the region where the real cross-section exists, are indicated. Different cross-sections are chosen in a consecutive order in this region. For each cross-section the data obtained from observations are computed. The discrepancy between the computed data and the observed ones is calculated. The set of cross-sections for which this discrepancy is sufficiently small is the solution of our problem. So the problem is to find the zone of minimum of a multidimensional function (the discrepancy between the computed and the observed data) in the space of the unknown parameters of the cross-section. In conclusion it is necessary to describe this set compactly, to show the common features of all found cross-sections.

*The method of direct inversion* has similar logics. We represent now through a finite set of parameters not the structure, but the curves, which we have to compare with observations: $t(\varDelta, h)$, $c(T)$ etc. The possible limits for each parameter, *i.e.* the region where the real curve lays, are indicated. Different sets of parameters are chosen inside this region. For each set the curves are computed. If the discrepancy between computed curves and observations is sufficiently small, the curve is inverted into structure; a set of all such structures is a solution of our problem.

The possibilities of this method are limited, since only the travel-time curve can always be inverted into structure (Sect. **3**); for dispersion curves, for example, inversion is possible, up to now, only when a good approximation to the structure is already known, so that a linear relation analogous to (30) holds.

That is why the *combined method* may be also useful [11]: to invert some kind of the data by the direct inversion method, and then to test the obtained structure by the trial-and-error technique against the rest of the data.

Each operation of the methods so described is connected with specific problems; some of them are still unsolved. We shall describe these operations in more detail, starting from the trial-and-error method.

5˙2. *The summary of observations.* – The raw observations should be used here *i.e.* the clouds of measured dots *without* drawing the above mentioned functions characterizing the wave-propagation.

For example, the arrival times should be given as a cloud of dots in the $(t, \Delta)$ plane for each $h$, without drawing the curve $t(\Delta)$, even less its loops; also surface wave velocities $c$ or $U$ should be given as a cloud of dots in the $(c, T)$ or $(U, T)$ planes, without drawing the dispersion curve. The error should be estimated for each dot. Only the simplest corrections, which make the dots comparable, are allowed here.

The dots should represent measurements which do not predetermine each other. In particular, the interval of $T$, for which $c$ and/or $U$ are given, should not be too small, as well as the intervals between eigenperiods $T_k$. The problem—on which intervals the measurements of $c$, $U$ or $T_k$ are still not completely correlated—is not solved.

The identification of body wave type, or of mode number, is not necessary and not even desirable as it is sometimes a powerful source of errors.

The waves should only be divided into the classes-body, Rayleigh or Love waves; the free oscillations—into spheroidal and torsional modes. With reliable data on polarization (or perhaps the spectrum) the body wave arrivals can be identified as longitudinal or transverse.

5˙3. *The summary of a priori data* considers the limitations on crosssection following from previous investigations and physical considerations, especially from equations of state. This summary should include possible relations between different properties: elasticity, density, attenuation, etc.

5˙4. *The parametrization of the cross-section* involves the representation of the cross-section through a finite number of numerical parameters and the indication of *a priori* limits for them. Those parameters are the parameters of the analytical approximation of $\alpha, \beta, \varrho, Q$.

It is, practically, impossible to represent these functions analytically in the whole depth-interval; and in any case they may be discontinuous. We, therefore, have to divide the structure into some depth intervals (layers) and separately approximate each physical function in each layer.

The parameters of the structure will be the parameters of the approximating functions in each layer as well as the depth of the layer boundaries. Usually the physical function inside each layer is approximated by a constant, a straight line, a parabola or some higher polynomial in $r$.

Few words on the choice of parameters. It may be decisive for the success of inversion. We have to avoid two extremities: a too simple and a too complicated approximation of the structure.

Evidently, it is important to keep the number of parameters $N$ as small as possible. If each unknown parameter can be equal to $K$ different values, then the total number of possible structures will be $K^N$. For the realistic situation $K = 8$, $N = 20$, we will have about $10^{18}$ possible structures, which is far from the nature of our problem.

Evidently, it is desirable to divide parameters into separate groups so that various data depend on various groups. Then one can find the minimum in each group independently. Obviously, this is equivalent to a reduction in the number of parameters. For example if two groups of $n$ and $N - n$ parameters are in such way decoupled, then the number of knots of the net will be $K^n + K^{N-n} \ll K^N$.

On the other hand, if the number of parameters is too small, we obtain a too rough approximation of the structure and possibly do not obtain at all its interesting elements. Thus, if $u(r)$ is approximated by the two straight lines, it is impossible to obtain a nonsurface wave-guide.

Evidently, the number of parameters should be sufficient to permit the existence of all the elements of the structure which may explain the main peculiarities of the given observations. For example, an abrupt increase of velocity-gradient has to be allowed if there is a loop at the travel-time curve; both the low-velocity layer and the zone of small velocity-gradient have to be allowed if a shadow is outlined, etc.

The proper choice of parameters can reduce their necessary number significantly. As a trivial example the linear approximation of a structure is much more economical than the approximation by homogeneous layers. That is why a large variety of parameters is usually used, such as: the depth of a boundary; the thickness of a layer; the value and/or the gradient of $\alpha$, $\beta$, $\varrho$ or $Q$ on each boundary; the increment, ratio or average value of all the parameters mentioned above.

The choice of parameters must be correlated with the physical task of inversion. If, for example, we seek to find out whether some elements of the structure exist (a wave-guide, or a discontinuity of velocity within some depth interval, etc.), then the assumed system of parameters has to allow the structures both with these elements and without them. The probability to meet the considered element by random choice should be about $\frac{1}{2}$.

Finally, the information, contained in the given observations, is the decisive factor in parametrization. This information can be estimated by the methods, described in Sect. **4**. More detailed description of the problems, connected with parametrization, can be found in [1, 10-13, 19].

As a result of this operation *we represent the cross-section as a point in the space of its unknown parameters,* and indicate their limits, thus determining the region in which the point lays. Limits can also be imposed on the values of parameters and on any function of them—the ratio, difference etc. [31, 20].

Our problem is to narrow this region as far as the observations allow.

5˙5. *The flow of cross-sections.* – Any measure of the discrepancy between observations and the computed properties of seismic waves depends on the parameters of the structure, because the computed properties depend on them. We have to find a set of combinations of the unknown parameters (*i.e.* a volume in the space of unknown parameters) for which the discrepancy is sufficiently small.

It is a wide-developed problem of computational mathematics, though usually not the region of the minimum but only the minimum itself is sought. There is no general theory of solution of this problem and the method of search of the minimum is to be chosen empirically.

The simplest and the only absolutely reliable method is to divide the investigated multidimensional region by a net, to calculate the function in each knot of the net and to choose the points in which the function is sufficiently small. The simplest method of construction of this net is to place the knots along co-ordinate axes at equal intervals. The step of the net must have a magnitude of the order of the error that is allowed for the determination of the boundaries of our minimum volume. Certainly, this error has to be chosen in accordance with the exactness of observations and computations: the difference of the minimized function at the neighbouring knots must be significant.

If the number of parameters $N$ is large (more than four), the number of knots $K^N$ becomes so great that the described method is, practically, impossible. We can construct a net by the *Monte Carlo method, i.e.* taking random points as knots. The number of knots of this net can be much less than in the previous case. However, it must also increase with the increasing of $N$, otherwise, the knots would be distributed unevenly and the distance between them would be too large, so that we might miss a part of the investigated volume.

Therefore, the *guided methods* for the search of the minimum are used if $N$ is large. Every next point in which the function will be calculated is chosen on the basis of the previous calculations, by some extrapolation of function. The reliability of the method depends on the reliability of the extrapolation.

Guided methods are divided into deterministic and random. In the deterministic methods, each next point is chosen only by one unique way. These methods are: the Gauss-Zaidel method; the method of quadratic minimization [14]; the method of steepest descent; the method of ravines; the iterative method in which the approximation at every step is carried out by the method of least squares.

In random methods the next point is chosen randomly corresponding to some distribution function obtained on the basis of the previous computations.

The points obtained by the random method are often used as the starting points for further descent by some deterministic method.

A description of these methods and the comparison of their efficiency in the inversion-of-travel time and amplitude data for body waves can be found in [1, 11].

The method of descent along a gradient is only efficient for a small number of unknown parameters $N$. For large $N$, the method of random search is efficient far from the minimum; but the search gives way to casual wandering in the vicinity of the minimum.

Strange as it may seem the Monte Carlo method proved to be preferable; however, it also possesses a great disadvantage in that the results of the trials already made are not used in the next trial. This is why only a small precentage of trials prove to be successful.

The most efficient method seems to be the « hedgehog method » [20, 12] which procedes as follows: a single point of the minimum region $y(y_1, ..., y_n)$ —the $y_i$ being the parameters of the cross-section—is found by one of the random methods—Monte Carlo or some other; and then the neighbouring points $y + \varepsilon_i dy_i$ are tried where $\varepsilon_i = 0$ or $\pm 1$; different combinations of the $\varepsilon_i$ are picked in turn. Points falling within this minimum region are then selected and the same procedure applied to these points until the whole region is covered. When this has been done we return to the random method (omitting, of course, the region just found from further investigation) and look for another minimum region, and so on.

The most direct approach is to test all combinations of $\varepsilon_i$ i.e. to test all the neighbouring points which can be obtained from an initial point by not more than one step along each co-ordinate axis. This is however computationally difficult for large $N$: each initial point has $(3^N - 1)$ such neighbours. For this reason the method includes the possibility of making the limitation that only those neighbouring points are tested for which no more than $k$ ($k < N$) of the coefficients $\varepsilon_i$ are nonzero. Thus, only those points which can be generated by taking one step along no more than $k$ co-ordinate axes from the initial point are tested. It is useful to look at the formulae for the number of neighbours.

Let $M(N, k)$ and $M^*(N, k)$ be the number of neighbouring points in an

$N$-dimensional grid, which have no more than $k$ nonzero $\varepsilon_i$ and exactly $k$ nonzero $\varepsilon_i$ respectively.

Then [15]

$$M(N, 1) = M^*(N, 1) = 2N ,$$

$$M(N, 2) = 2N^2 ,$$

$$M(N, N) = 3^N - 1 ,$$

$$M(N, k) = M(N, k-1) + M^*(N, k) ,$$

$$M^*(N, k) = 2[M^*(N-1\ k-1) + M^*(N-2, k-1) + \dots M^*(k-1, k-1)] .$$

Figure 10 illustrates the hendgehog method for a two-dimensional space $(N=2)$. Heavy lines represent a grid for which the horizontal increment is twice larger than for thin lines. The true area of the minima consists of two closed parts, $A$ and $B$. Suppose we found point 1 by a Monte-Carlo technique; assume $k=1$; then we should try the neighbouring points 2-5. Point 2 belongs to our minimum region; so we then try points 6-8 around it.

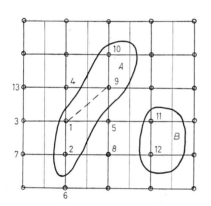

Fig. 10. – Explanation of the hedgehog method. Circles: points of the rare grid (heavy lines); thin lines: finer grid. $A, B$: the true regions of minima. After [20].

To find the rest of our minima we have to return to a random search until we find the points 9 or 10 and 11 or 12. However, we would think of points 1, 2 as belonging to a region isolated from the points 9, 10 until we check it by taking small steps along the dashed path or introducing a finer grid.

For $k = N = 2$ we would, however, try at once the points 2-5, and 7-9, 13 around point 1, and then point 11 would be found from 9; in this case we would think of $A$, $B$ as a single area until the finer grid would be introduced.

This example shows that if the increments $\varepsilon_i$ are too big or $k$ is too small, we can, easily, miss some points of the area of minima or introduce a ficticious division of this area into isolated parts. However, decreasing $\varepsilon_i$ and increasing $k$ leads to a large increase in the number of computations. Decreasing $\varepsilon_i$ is particularly inconvenient: it gives a lot of internal points in the area of minima which are of no interest. It is better to increase $k$ or preferably to rotate the co-ordinate axes.

**5·6.** *The direct-problem computation.* – The theoretical computation of the data which are known from observations needs no comment. Practically, it can be done for any kind of data if the model of the structure is horizontally homogeneous.

For horizontally inhomogeneous models the computations are comparatively well-developed only for body waves in the ray approximation.

**5·7.** *The discrepancy between computations and observations.* – The choice of the measure of this discrepancy is a most important but inadequately developed problem. This measure should, ideally, be very dependent on those properties of the cross-section we wish to investigate, and depend little on other properties; unfortunately the task of inversion is not always specified in advance.

In practice, the standard and maximal deviations between observations and computations are used as the measures of their discrepancy; taking them separately for each kind of observation.

A structure is rejected if at least one of these measures is greater than the corresponding threshold. The definition of these measures, and even more that of the thresholds is essentially intuitive; we have to rely here on common sense, or (Prof. CAPUTO) on « faith and imagination », or perhaps—on lack of imagination. Practically, the uncertainty can be resolved at the present time by assuming sufficiently large thresholds. This means that in the absence of strict theory we must reject only those structures which show a large discrepancy between observations and calculations.

This situation is evidently unsatisfactory: if seismology is a branch of physics, we have to formulate our results in terms of a confidence level, *i.e.* to determine a set of structures which contains the true one with a given probability.

At the present moment the only problem, for which a complete solution of such a kind is obtained in seismology is the much simpler problem of the determination of confidence limits for the fault-plane orientation [4]. Two methods, however, are in the stage of development [15, 19, 25] and I will describe them schematically.

Denote by $x$ the vector of the observations, by $\theta$ the vector of the parameters of structure, and by $L(x/\theta)$ the density of the probability distribution of observing $x$ if the « true » model is determined by $\theta$. All three sources of errors—from observations, idealization and computations—should be included in such a distribution. Let us compute $L(x/\theta)$ for a given $x$ and all possible $\theta$; both $x$ and $\theta$ are multidimensional, but the idea can be understood from the one-dimensional scheme in Fig. 11. This Figure shows a plot of the product $L(x/\theta) P(\theta)$, where $P(\theta)$ is the *a priori* probability of $\theta$. The interval in $\theta$, which contains $Q\%$ of the total area of the curve $LP$ contains the true solution with probability $Q\%$. The multidimensional generalization is immediate.

Another method which needs no knowledge of the *a priori* distribution $P(\theta)$ is the following [19]: denote by $P(x/\theta)$ the distribution function corresponding to the density of the distribution $L(x/\theta)$; let us try all $\theta$ (in some discretion, of course) and select a set of $\theta$ for which $P(x/\theta) \geqslant (100 - Q)\%$. Then the probability of not selecting the true $\theta$ is $(100 - Q)\%$; if the true $\theta$ is included in those tested, as has been assumed, our set contains a true $\theta$ with probability $Q\%$.

There are three major difficulties here: the definition of $L(x/\theta)$, $P(\theta)$ and of $P(x/\theta)$ even for known $L(x/\theta)$—the last difficulty arising from the multidimensional nature of $x, \theta$.

The model of a likelihood function for multi-valued travel times was suggested in [15], and developed further in [19, 25]. The statistical problem as a whole remains unsolved.

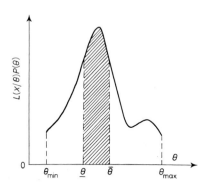

Fig. 11. – Scheme of the likelihood method. $0_{\text{min}}$, $\theta_{\text{max}}$—*a priori* limits for $\theta$, $\underline{\theta} < \theta < \bar{\theta}$—confidence interval for $\theta$; its level is equal to the darkened area divided by $\int_{\theta_{\text{min}}}^{\theta_{\text{max}}} L(\theta/x) P(\theta)\, \mathrm{d}\theta$.

We have to describe three more operations, involved into the direct inversion method (Fig. 8 $b$)).

**5ˈ8. *The parametrization of curves*** means the representation of each curve by a finite number of parameters and the indication of *a priori* limits for them; this operation being based on the general theory of these curves, on *a priori* data and on the pattern of observational dots. The general theoretical properties of the travel-time curves are given in [5, 16], but they have not yet been formulated for dispersion curves.

The problem of optimal parametrization of curves is equivalent to that for cross-sections and is not solved yet. In routine practice we represent each curve by some set of ordinates or by the parameters of straight lines and parabolae which approximate separate pieces of the curve.

**5ˈ9. *The flow of curves*** means here the sequential picking of various combinations of the corresponding parameters; considering, of course, all possible variants of wave identification, for each variant of the curve.

It is very convenient that the flow of curves can be organized on a two-dimensional plane *e.g.* $(t, \varDelta)$, $(\mathrm{d}t/\mathrm{d}\varDelta, \varDelta)$, $(c, tT)$; $(U, T)$, etc.

5˙10. *The inversion of the curve* into a cross-section (or the parameters of the curve into parameters of the cross-section) is the next operation, and is applied to all variants of the curve which fit the observations.

At the present time we can invert only the data on travel-times $t(\Delta)$ and/or on $dt/d\Delta$ (see Sect. **1**); for dispersion curves and free oscillations it is possible only in a linear approximation of the type (30).

## 6. – Some examples of inversion of real data.

Figure 12 shows what is apparently the first attempt to get a complete set of solutions corresponding to given observations [22]. It shows some of the 115 structures of the upper mantle, corresponding to *P*-wave travel-times in Europe, after [22], [23].

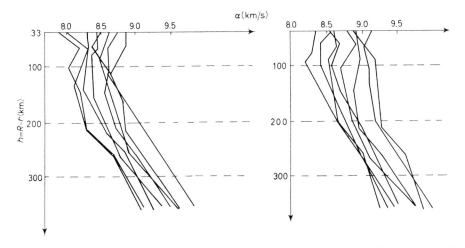

Fig. 12. – Some of the 115 longitudinal wave velocity distributions of the upper mantle, corresponding to the travel-times of *P*-waves in Europe (after [22, 23]). The horizontal scale is given for the first left curve. Each following curve is shifted to the right in relation to the preceding one by 0.1 km/s.

The horizontal scale corresponds to the left-hand curve, each of the others being shifted to the right successively.

Figure 13 shows the usefulness of the joint inversion of different data [10]: it shows *S*-wave velocity distributions corresponding to the *S*-wave travel-time curve and Love-wave dispersion for the Canadian Shield: the heavy line is the structure which corresponds to both kinds of data. The result, however, should be treated with some reservations [10]: it is rather an example of a method. More definite results of joint inversion of body and surface waves

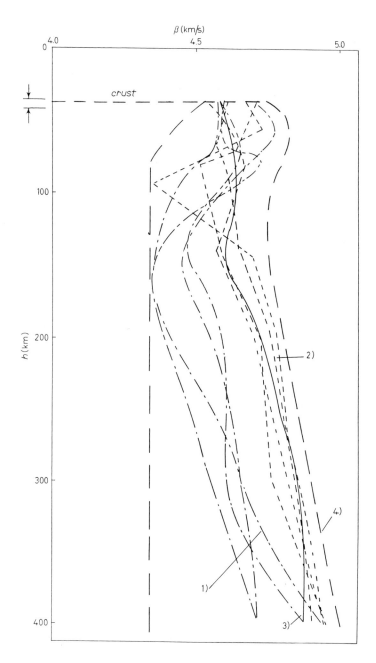

Fig. 13. – Velocity distributions corresponding to the observed *S*-wave travel-time (1) and the observed dispersion of the fundamental mode of Love waves (2); phase velocities for (10÷70) s, group velocities for (10÷140) s. (3) a common velocity distribution; (4) the *a priori* limits of variation of velocity. After [10]. —·—·— 1), — — — 2), ——— 3), — — — 4).

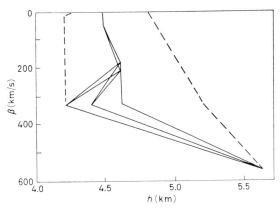

Fig. 14. – Shear-wave velocity distributions in Europe. Data used: travel-times of $S$-waves (first arrivals) at the distances up to $40°$; amplitude-distance curves for the same distances; Love- and Rayleigh-wave dispersion of the periods $(20 \div 60)$ s. After [12].

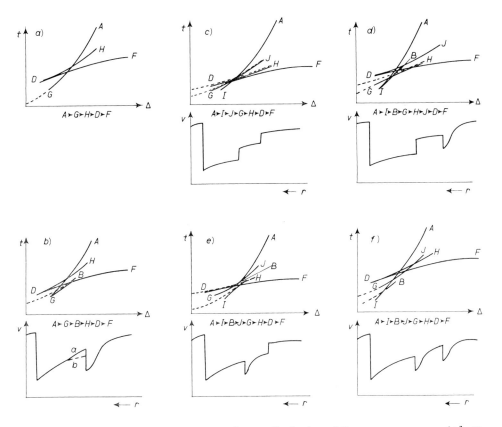

Fig. 15. – An example of direct inversion method: travel-time curves, generated on computer (on $(\mathrm{d}t/\mathrm{d}\Delta, \Delta)$ plane) and fitting PKP data on $120° \leqslant \Delta \leqslant 160°$. The types of structure are shown below each travel-time curve. $a)$, $b)$: two-layered core; $c)$, $d)$, $e)$, $f)$: three-layered core. The sequence of branches of $t(\Delta)$, corresponding to decrease of $\mathrm{d}t/\mathrm{d}\Delta$, is indicated by the sequence of letters. After [25].

were obtained in [12] for the upper mantle in Europe and for the Earth's crust
in one of the regions of Central Asia [21]. Figure 14 shows the $S$-wave velocity
distribution in Europe, after [12]. A further important widening of the data
inverted was the incorporation of the eigenperiods and gravity data, within
the same methodical framework, as shown on Fig. 9 [17, 18]; the results and
references can be found in [18].

Figure 15 shows an example of the method of direct inversion: some of the
structures obtained by the search on the $(t, \Delta)$ plane, for a transition zone between the outer and inner core, after [25].

An international group of scientists from six countries (Canada, India, Israel,
New-Zeland, USA, USSR), working at the University of California, Los Angeles, has constructed a package of programs for inversion of an arbitrary set
of data, (this work was a part of the International Upper Mantle Project).

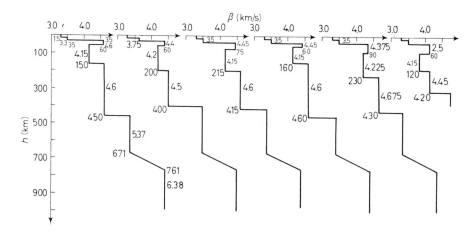

Fig. 16. – Shear-wave velocity-distributions for the Western United States, up to
$h = 400$ km. Data used: travel-times, amplitudes and $dt/d\Delta$ for $S$-waves at $3° < \Delta < 60°$
(up to three first arrivals); Lowe-wave and Rayleigh-wave dispersion in the period
range $(40 \div 220)$ s. After [13].

The brief description of the method and first results are given in [13].
Figure 16 shows the transverse-wave velocity-distribution, obtained in [13]
from the data, related to the Western United States: Love and Rayleigh-
wave dispersion; travel-time $t(\Delta)$, direct measurements of $dt/d\Delta$ and amplitudes
of some arrivals.

It is important to note that the existence of wave-guides in this case is
controlled by the combination of only two simple samples of data: the $S_n$
velocity, and the surface-wave velocity in the period range from 50 to 120 s.
The first gives the velocity of shear waves at the very top of the mantle while

the second averages the shear wave velocity of the top $(200 \div 300)$ km of the mantle. If the first is larger than the second, then a waveguide exists.

This example gives a hope that the careful preliminary analysis of data will make the inverse problem not so difficult after all.

\* \* \*

I am very grateful to Mr. B. L. N. KENNET of the Department of Applied Mathematics and Theoretical Physics at the Cambridge University, for the thoroughful revision of this paper.

## REFERENCES

[1] J. AZBEL, V. KEILIS-BOROK and T. JANOVASKAYA: Geophys. Journ. Roy. Astr. Soc., 11, 25 (1966).

[2] J. BACKUS and F. GILBERT: Geophys. Journ. Roy. Astr. Soc., 13, 247 (1967); 16, 169 (1968); Phil. Trans. Roy. Soc., A 266, 123 (1970).

[3] E. N. BESSONOVA, G. A. SITNIKOVA and V. M. FISHMAN: Determination of the limits of averaged velocity-distribution from the final set of observations on body-waves (in Russian), Theoretical and Computational Geophysics, Proceedings of the Soviet Geophysical Committee (in press).

[4] T. GELANKINA, V. KEILIS-BOROK, V. PISARENKO and J. PJATEZKI-SHAPIRO: Computer determination of a fault plane (in Russian); Computational Seismology, Vol. 5 (Vichislitelnaya Seixmologiya, Vol. 5) (1970).

[5] M. L. GERVER and V. M. MARKUSHEVICH: Geophys. Journ. Roy. Astr. Soc., 11, 165 (1966).

[6] M. L. GERVER and V. M. MARKUSHEVICH: Geophys. Journ. Roy. Astr. Soc., 13, 241 (1967).

[7] M. L. GERVER: Geophys. Journ. Roy. Astr. Soc., 21, 337 (1970).

[8] I. O. NUTTLI: Travel-time curves and upper mantle structure from long period S-waves, BSSA (1967).

[9] H. JEFFREYS: The Earth, V edition (Cambridge, 1970).

[10] V. J. KEILIS-BOROK: Seismology and Logics, Research in Geophysics, Vol. 2 (M.I.T., 1964).

[11] V. KEILIS-BOROK and T. YANOVSKAYA: Geophys. Journ. Roy. Astr. Soc., 13 233 (1967).

[12] V. KEILIS-BOROK, A. LEVSHIN and V. VALUS: S-wave velocities in the upper mantle in Europe, report on the IV International Symposium on Geophysical Theory and Computers 1967, Dokladi Akademii Nauk SSSR 185 (1968), No. 3, p. 564.

[13] V. KEILIS-BOROK and L. KNOPOFF: report on the VI Symposium on Geophysical Theory and Computers, UMC (Copenhagen, 1969), (in print).

[14] L. KNOPOFF: Suppl. Nuovo Cimento, 6, 120 (1968).

[15] L. KNOPOFF and T. L. TENG: Rev. Geophys., 3, 11 (1965).

[16] V. M. MARKUSHEVICH: Properties of travel-time curve from deep focus (in Russian),

*Computational Seismology*, Vol. **4** (Vichslitelnaya Seismologiya, Vol. **4**) (1968), p. 64.

[17]  F. PRESS: this volume p. 209.

[18]  F. PRESS and S. BICHLER: *Journ. Geophys. Res.*, **69**, 2979 (1964).

[19]  M. SCHNIRMAN and T. YANOVSKAYA: *On the likelihood function for travel-times, theoretical and computational geophysics*, in *Proceedings of the Geophysical Committee, Ac. Sci. USSR* (in Russian) (in print).

[20]  V. VALUS: *Determination of seismic structures from the set of observations* (in Russian), *Computational Seismology*, Vol. **4** (Vichislitelnaya Seismologiya, Vol. **4**).

[21]  V. VALUS, A. LEVSHIN and T. SABITOVA: *Joint interpretation of body- and surface-wave data for the one of the regions of Central Asia* (in Russian), *Computational Seismology*, Vol. **2** (Vichislitelnaya Seismologiya, Vol. **2**) (1966).

[22]  T. B. JANOVSKAYA: *Computation of velocity-depth curves for the Upper Mantle in Europe as the inverse mathematical problem*, Veröffentlichungen des Instituts für Bodendynamik und Erdbebenforschung in Jena, Heft 77 (1962).

[23]  T. B. JANOVSKAYA: *Determination of velocity-distribution in the Upper Mantle, as an inverse mathematical problem*, Izv. Akad. Nauk. SSSR, Ser. Geofiz., No. 8 (1963).

[24]  T. B. JANOVSKAYA and I. Y. ABSEL: *Geophys. Journ. Roy. Astr. Soc.*, **8**, 313 (1964).

[25]  T. B. JANOVSKAYA: *Geophys. Journ. Roy. Astr. Soc.* (1968).

# On Free Periods of the Earth above One Hour.

C. Denis

*Laboratoires de Géophysique, Université de Liège - Liège*

It is commonly believed that the slowest possible free mode of the Earth is $_0S_2$ with a period of about 53.82 min [1]. However, quite a few unforeseen peaks with periods ranging from 61 to 152 min (at least) appear in many power spectra obtained after large earthquakes. Some of these peaks are significant at a confidence level higher than 95 %. Most probably, all the concerned motions are of the spheroidal type; they do not seem to be explainable as indirect or as local effects. The geographical position of the epicentre is apparently irrelevant as regards their occurrence.

No fully convincing theory that would predict these peaks correctly has been put forward yet. However, SLICHTER [2] considers rigid body motion of the solid inner core in the surrounding fluid outer core as a possible explanation of a particular 86 min gravimeter observation, but this interpretation remains very hypothetical for several reasons, one of which is that an exceedingly high inner core density is needed. CALOI ([3], pp. 183-189) supposes that the crust and part of the upper mantle form a spherical shell which is elastically uncoupled from the lower regions of the Earth by the asthenosphere and which gives rise to independent free elastic oscillations; the latter would have about the correct periods to explain some of the unidentified peaks, but many points in this theory remain obscure. CAPUTO and his co-workers [4, 9] interpret the motions with periods above one hour as combinatory modes resulting from nonlinear interaction between particular normal modes. This theory does not appear quite satisfactory neither, although bispectral analysis gives some support to it. In my opinion all the observed peaks should be predictable by a linear theory; the very long periods indicate that gravitational rather than elastic restoring forces are acting.

Investigations on stellar oscillations [5] have shown that in the case of fluid spheres four independent spectra of spheroidal eigenfunctions are generally associated to a given spherical surface harmonic: these are called $p$-, $f$-, $^+g$- and $^-g$-spectra by astrophysicists.

The infinite discrete $p$-spectrum is quite similar to the usual seismological $_nS_1$-spectrum. The $f$-spectrum consists of a single mode, with eigenperiod higher than the fundamental period of the corresponding $p$-modes; in fact, for a homogeneous sphere the $f$-mode reduces to the well-known Kelvin mode. The infinite discrete $^+g$- and $^-g$-spectra arise when gravity can act as the dominant restoring (or repelling) force. This happens in regions where the barotropic relation between pressure and density is not adiabatic. Indeed, the existence of $g$-spectra (« $g$ » for gravitational) is closely tied up with the value of the difference between the actual and the adiabatic temperature gradients

$$A = \frac{1}{\varrho} \frac{\mathrm{d}\varrho}{\mathrm{d}r} - \frac{1}{\varkappa} \frac{\mathrm{d}p}{\mathrm{d}r},$$

$\varrho$ is the specific mass, $p$ the pressure, $\varkappa$ the adiabatic bulk modulus, and $r$ is the central distance of any spherical shell.

If $A > 0$ throughout the self-gravitating sphere, no $^+g$-spectrum exists, but an infinite discrete set of negative eigenvalues appear which accumulate towards zero (i.e., the corresponding time scales increase towards infinity); these lead to a dynamical instability in the associated eigensolutions, which LEDOUX [6] has related to local convective instability as indicated by Schwarzschild's criterion. Therefore, the $^-g$-modes are often called convective or unstable gravitational modes.

If $A < 0$ everywhere inside the sphere, no $^-g$-spectrum occurs, but an infinite discrete $^+g$-spectrum does. All the eigenvalues are strictly positive [7] and accumulate towards $0^+$, that is, the fundamental $^+g$-eigenperiod is higher than the $f$-period and the $^+g$-overtone periods increase steadily with the third mode number.

If the sphere has a completely adiabatic structure ($A = 0$ everywhere) both $g$-spectra degenerate to frequency zero, which means neutral stability.

If the quantity $A$ takes positive as well as negative values inside the sphere, both types of $g$-spectra occur simultaneously. However, in stable ($A < 0$) regions, the $^+g$-eigenfunctions generally oscillate, whereas the $^-g$-eigenfunctions decrease exponentially; in unstable ($A > 0$) regions, the converse situation holds.

This shows the possibility of obtaining ultraslow free vibrations in fluid spheres, and the essential role played by the quantity $A$. This role has not yet been investigated explicitly when rigidity enters the problem, as far as I know. LOVE [1], when discussing the free oscillations of a spherical earth model with uniform density and uniform elastic parameters (for which $A$ is of course strictly positive everywhere), distinguished three types of spheroidal motions: quick vibrations, an intermediate vibration, and slow vibrations. The quick vibrations compose the usual $_nS_1$-spectrum, but the physical nature of the slower vibrations is not quite clear. LOVE found that slow vibrations cannot occur if the rigidity of the body exceeds some critical value. On physical grounds I would suggest that Love's vibrations of the slow type are unstable gravitational modes which have been stabilized by rigidity; of course, such an inference needs a mathematical confirmation. An attack of this problem is under progress. Nothing is known when $A < 0$ and rigidity is small.

Anyhow, the former discussion seems to indicate that free vibrations with periods over an hour do exist and correspond to gravitational restoring forces, provided there are regions in the Earth with small or no rigidity possessing a structure for which the Adams-Williamson equation does not hold. However, several estimates [8] point out that the upper mantle, in particular the low-velocity layer, is such a region. If so, the peaks with periods above one hour deserve much more attention than they have received until now. Indeed, their study should become a promising tool for investigating thermal structure; also, the present gap of about an order of magnitude between the longest periods of free oscillations and the periods characteristic of the earth tides would be reduced considerably, so that dynamic effects, maybe, should be accounted for.

* * *

An expanded version of this note will be published elsewhere. I wish to thank Prof. J. COULOMB, Director of the Summer Course, and the Italian Physical Society for giving me the opportunity to speak about this problem, and Prof. M. CAPUTO for drawing my attention during the course to his work on interaction modes.

## REFERENCES

[1]  A. E. H. LOVE: *Some Problems of Geodynamics* (Cambridge, 1911), p. 89 (Dover reprint, 1967).
[2]  L. B. SLICHTER: *Proc. Natl. Acad. Sci. Wash.*, **47**, 186 (1961).
[3]  P. CALOI: *Adv. Geophys.*, **12**, 79 (1967).
[4]  M. BOZZI ZADRO and M. CAPUTO: *Suppl. Nuovo Cimento*, **6**, 67 (1968).
[5]  P. SMEYERS: *Bull. Soc. Roy. Sci. Liège*, **36**, 357 (1967).
[6]  P. LEDOUX: *Mém. Soc. Roy. Sci. Liège*, 4ª série, **9**, 172 (1949).
[7]  N. R. LEBOVITZ: *Astrophys. Journ.*, **142**, 229 (1965).
[8]  W. M. KAULA: *An Introduction to the Planetary Physics: The Terrestrial Planets* (New York, 1968), p. 114.
[9]  M. CAPUTO and F. ROSSI TESI: *Suppl. Nuovo Cimento*, **6**, 857 (1968).

# PROCEEDINGS OF THE INTERNATIONAL SCHOOL OF PHYSICS
## « ENRICO FERMI »

*Information about Courses I-XIII may be obtained from the Italian Physical Society.*

Tipografia Compositori - Bologna - Italy

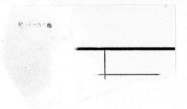